THE ANATOMY OF MARTIAL ARTS

AN ILLUSTRATED GUIDE TO THE MUSCLES USED IN KEY KICKS, STRIKES & THROWS

Dr. Norman Link and Lily Chou

Illustrations by SUMAN KASTURIA

Ulysses Press

This book is dedicated with humble appreciation to Dr. Ken Kyungho Min, founder of the University of California at Berkeley's Martial Arts Program (UCMAP). It was Dr. Min's encouragement to explore all aspects of martial arts that made this book possible.

Published in the United States by
ULYSSES PRESS
P.O. Box 3440
Berkeley, CA 94703
www.ulyssespress.com

Library of Congress Control Number 2009940338
ISBN13 978-1-56975-787-1

Printed in the United States by Bang Printing

10 9 8 7 6 5 4 3 2

Acquisitions Editor: Keith Riegert
Managing Editor: Claire Chun
Editor: Lauren Harrison
Proofreader: Bill Cassel
Interior design and layout: what!design @ whatweb.com
Front cover design: Double R Design
Back cover design: what!design @ whatweb.com
Production: Judith Metzener
Anatomy illustrations: Suman Kasturia except on page 136 © ANATOMIA3D/shutterstock.com
Photographs: Rapt Productions
Models: Jon Bertsch, Lily Chou, David Commins, Luke Commins, Kelly Kim, Norman Link, Susan Link,
    Bob Matsueda
Index: Sayre Van Young

Distributed by Publishers Group West

Please Note:
This book has been written and published strictly for informational purposes, and in no way should be used as a substitute for consultation with health care professionals. You should not consider educational material herein to be the practice of medicine or to replace consultation with a physician or other medical practitioner. The author and publisher are providing you with information in this work so that you can have the knowledge and can choose, at your own risk, to act on that knowledge. The author and publisher also urge all readers to be aware of their health status and to consult health care professionals before beginning any health program, including changes in dietary habits.

Table of Contents

PART 1: OVERVIEW

INTRODUCTION

Welcome to *The Anatomy of Martial Arts: An Illustrated Guide to the Muscles Used in Key Kicks, Strikes and Throws*. Between the two authors, we have about 60 years of formal martial arts training and yet are just starting to scratch the surface of learning. This is not an attempt at being humble; it's a simple fact. As you train in whatever martial arts you choose, your body changes. With luck, it is sculpted to flow with the techniques that the art demands, and with time there should be a steady improvement. However, when looking at martial arts training over a longer period of time, our bodies inevitably age and our physical abilities slowly decline. The bottom line is that we spend more and more of our time trying to adapt the techniques we know to an ever-changing set of bones and muscles.

For this book we've been limited to showing 50 techniques from as wide an array of martial arts as we could. Thus, we chose a number of hand strikes (including breaks), kicks, throws, weapon and grappling techniques, and rolls and falls. While a beginning martial arts student may find this book interesting, it will be most useful to intermediate and advanced practitioners of the martial arts.

Unlike most other martial arts books, this book assumes that the reader is already familiar with the techniques that are featured. We don't teach any techniques; rather, we highlight and discuss the main muscle groups required for the technique to be performed and suggest ways to both strengthen and stretch those muscles to improve the technique's quality. Because even basic moves such as a front kick can be taught a variety of ways depending on the art in which they're used, we hope that by emphasizing the body's fundamental structures, particularly the musculature and kinetic chains, the foundation of each technique might be reopened to discussion.

Even if you decide that the muscles highlighted are incorrect or incomplete, then at least we've accomplished our primary goal of getting you to think about each technique's foundation. We hope that by reviewing your movements as to which muscles are being used, you can augment your training to improve the power and motion that actually drive the techniques.

ANATOMY AND MARTIAL ARTS

Every move we make, be it sitting, standing, running, or kicking, involves an elaborate choreography of the 250 skeletal (or voluntary) muscles as they move our 206 bones. These bones are arranged as follows:

29 in the head and neck

2 clavicles, or collarbones (the most commonly broken bone in the body)

2 scapulae, or shoulder blades

26 in the spine, or vertebral column

24 ribs

1 sternum

2 in the pelvis

60 in the arms (3 each) and hands (27 each)

60 in the legs (4 each) and feet (26 each)

In brief, each muscle group has a specific set of functions and is often paired with an opposing muscle or muscle group. The biceps, for example, are responsible for bending the arm at the elbow, while the triceps are responsible for straightening it. Contracting the biceps causes the arm to bend; at the same time, the triceps must relax. Any disruption in this play of opposites can affect the movement (for example, tight biceps will prevent full arm extension). The last page of this book features a color-coded illustration of the muscles and their actions. You'll also find charts in the appendix that list the key muscles and their functions.

The Anatomy of Martial Arts largely ignores the 29 bones in the head, except insofar as to recognize that the head must be protected (as with a chin tuck during a back fall). The movements of the remaining 177 bones and the muscles that move them are what make the practice of martial arts so very interesting and difficult to learn. The martial arts, when properly performed, aren't just a set of actions but a veritable symphony of movements. This makes identifying the muscles involved in any given technique a challenge. Even a technique as seemingly simple as a reverse punch requires the martial artist to perform a specific sequence of actions in a specific order and with specific timing.

It's beyond the scope of this book to describe all the muscles involved in each stage of a technique; rather, this book highlights the key muscles and the kinetic groups they work in. We hope that this will help you reconsider how you perceive the various techniques and how you might improve them.

LINES OF POWER FOR MOVEMENT: KINETIC CHAINS

Power is required not only for hand strikes and kicks but also for throws, jumps, falls, and twisting out of an attacker's reach. A number of people have used the term "kinetic chain" in reference to a power stroke of the body, or when muscles work together to produce a given line of power. While several kinetic chains have been defined and used in other works, this book features six major ones. (There are, of course, many others that can be defined, but for the sake of simplicity we'll stick with six.) With the remarkable complexity of even "simple" martial arts techniques, it's rare when there are not at least two of these kinetic chains working together to produce a flow of power in a desired direction.

The six kinetic chains described below are each responsible for a different key power drive of the body. Each description includes the relative effective range, speed, and strength, as well as a couple of examples of techniques that are based on that kinetic chain.

Posterior Kinetic Chain: This forward drive of the hips (sometimes referred to as a pelvic thrust) is a medium-range, slow, strong movement that's usually used to align the drive of the legs with either the weight of the torso or an upper-body drive. This kinetic chain is perhaps the hardest one to understand and is often a central component in *ki* exercises and other fundamental power-generation techniques. It gets its name from the fact that the muscles involved are on the posterior side of the body and range from the hamstrings in the legs all the way up to the latissimus dorsi in the upper back. It is essential in a standard reverse punch or a groundwork bridge.

Leg Extension Kinetic Chain: This long-range, fairly quick, strong drive involves the extension of the leg at the hip, knee, and ankle joints. It's usually associated with a kick or a lifting action of the body.

Hip Turn Kinetic Chain: This drive is short-range, slow, and very strong. The turn of the hip is intimately connected with leg movements and body twists, such as the sweeping hip throw.

Lateral Kinetic Chain: This medium-range, slow, mid-strength drive involves twisting the body to one side, such as with a side kick, some throws, and many ground techniques.

Shoulder Turn Kinetic Chain: This drive is short-range, medium speed, and strong. The turn of the shoulder is intimately connected with arm movements and, to a lesser extent, body twists. Hand strikes are common examples.

Arm Extension Kinetic Chain: This drive involves the extension of the arm at the shoulder, elbow, and wrist joints and is long-range, very quick, and medium strength. It's usually associated with a hand strike, block, or a pushing away of the body.

A strongly positioned base for each kinetic chain is critical for the efficient transfer of energy into an opponent. For example, a relaxed shoulder will result in the poor transfer of power during an arm extension such as a punch, while a solid pelvic girdle will result in a stronger, more effective kick. Thus, kinetic chains rely on muscle groups pushing against a firm part of the body or something solid, such as the ground.

Let's look at a simplified example of the many kinetic chains used in a right-hand reverse punch:

1. Step forward with your left leg, driving your body forward with your back (right) leg [*posterior kinetic chain*].

2. Stiffen your front leg (to create a pivot point) and, using your back leg and hips, twist your right hip forward [*hip turn kinetic chain*].

3. Using the stiffening muscles of your legs, hips, and torso as a base, twist your shoulders to drive your right shoulder forward and your left shoulder back [*shoulder turn kinetic chain*].

4. Using the now-stiff muscles as a base, straighten and twist your right arm to deliver the punch [*arm extension kinetic chain*]. Note that turning the palm of your right fist down effectively twists the two bones of the forearm (ulna and radius) together to make a stiffer arm, which is more efficient at transferring the strike's energy to the target.

While the above is obviously oversimplified and incomplete, it illustrates the idea that even a "simple" reverse punch is the result of a complex and well-coordinated sequence of actions. This mixture of using both dynamic (moving) and static (tensed but not moving) muscles makes the timing and, thus, the description of the various techniques illustrated in this book very difficult. However, attempting to break down these techniques into their component parts allows us to suggest various exercises and stretches to further strengthen the moves.

Let's look at a much more complex example of the kinetic chains used in a butterfly kick:

1. From a standing position, turn sharply to the left and step out and back as your arms extend and your body dips parallel to the floor [*shoulder turn, hip turn, lateral, and arm extension kinetic chains*].

2. Bend your left leg and continue to drive your body down and around to gather momentum.

3. Extend your left leg to drive your body into the air as your straight right leg and arms arch behind your back [*leg extension kinetic chain*].

4. Open your body flat for the middle part of the flight [*posterior kinetic chain*].

5. Pull your right leg down and forward to catch your body weight as you land.

Butterfly kick

CONSEQUENCE OF IMPACT AND MISUSE

Martial arts in general involve a certain amount of impact to the body. Most of the impacts are obvious, such as kicks and punches in the striking arts and falls that are taken in the throwing arts. One of the most dangerous and least understood results of impact in any sport is a concussion or bruising of the brain, which can be caused by blows to or violent shaking of the head and neck; these must be taken very seriously as they can have both short- and long-term effects. Other impacts are not quite so obvious, such as the long-term effects of striking various body parts, including the hands and feet, against hard objects such as bricks and boards. Many don't recognize that while the short-term effects of such blows may be mild, the long-term effects (e.g., arthritis) can be serious and life-altering.

It's a staple of martial arts demonstrations to be struck in the abdomen with no ill effects. However, it's important to remember that being struck is inherently dangerous and must be done only under controlled circumstances; even the best-trained practitioners need a moment to tense their muscles so as to deflect the energy of the blow away from their vulnerable organs. When the world-famous magician Harry Houdini (1874–1926) was in his early 50s, he was still performing his physically demanding escapes and was by necessity in great physical shape. One of his demonstrations of his physical prowess was to invite big, strong, young men to punch him in the stomach. He did this repeatedly and suffered no ill effects from the blows. One day, a young man came into Houdini's dressing room and surprised him by punching him when he wasn't ready. Houdini died a few days later due to a ruptured bowel.

People in their first few decades of life who have decided to "toughen their limbs" should reconsider doing so—the damage they inflict on their bones and muscles may not start to severely impact their lives until they're in their 40s or 50s. Some of the more common long-term injury sites are the hands and feet as the result of striking hard targets, and the elbows and knees from repeated impacts, twistings, and hyperextensions. The latter injuries can be greatly exacerbated by the overuse of ankle and wrist weights.

Two other common muscle abuses include: (1) repeating an action until physical damage occurs (repetitive stress issue) and (2) so-called secondary injuries, which arise when an injured practitioner tries to continue training. The latter scenario results in the practitioner doing things in an awkward or imbalanced way. For example, if you injure your right knee, to avoid injuring it further you'll likely place additional stress on your left leg, resulting in a secondary injury due to this unbalanced practice. While from a practical point of view we understand that martial artists are often in a state of mild injury and that they must continue working out through these inconveniences, it must be done in an intelligent way so as to avoid incurring further injuries.

By learning and practicing proper martial arts techniques, the consequence of impacts on the body can be minimized and, within certain limits, martial arts can be practiced well into old age.

STEROIDS

The term "steroids" refers to a broad class of hormones. Some types of steroids, such as cortisone (only available by prescription and used to treat problems like asthma and arthritis), can be beneficial when used correctly. A group of artificial hormones called anabolic steroids comes in hundreds of varieties that are used to artificially enhance muscle mass, strength, and endurance. These illegal, testosterone-like hormones also cause numerous short- and long-term side effects, ranging from hair loss to heart disease to liver damage. While all the long-term issues that accompany taking anabolic steroids are not known, one thing has been well proven: The increase in muscle mass does *not* extend to the proportionate increased development of the bones and ligaments. Thus, the increase in muscle mass leads directly to irreversible joint and bone damage. It's our recommendation that steroids never be used unless prescribed by a doctor.

THE PHYSICS BEHIND A HIGH-ENERGY STRIKE

Martial arts practitioners commonly ask, "How can I get as much energy as possible into a strike?" The definitive answer is complex (think physics equations) and generally not very helpful. In addition, many factors are involved in generating a high-energy strike, including the relative velocity of the striking surface and the target, the elasticity of the striking surface (usually a hand or foot) and the targeted surface, body masses, etc. At the risk of oversimplifying the answer, we'll work with three relatively simple concepts.

Concept #1—Dynamic and Static Muscles: A *dynamic muscle* is defined as one that moves a part of the body; these are used to accelerate the body into a technique so that it has appropriate velocity. *Static muscles* are tensed but not moving, helping to put as much of a person's body mass behind a movement or blow as possible. Another way to think about this is to realize that many muscles work in opposition to others and, for a given action, one is the agonist, used for speeding the action, and the other is the antagonist, used for slowing it. For maximum velocity, the antagonist must relax when the agonist tenses, or contracts. For example, during a punch, the triceps (the agonist) extends the arm while the biceps (the antagonist) relaxes. However, at the end of the motion, it's usually recommended that the antagonists be used to slow the movement in a controlled manner as opposed to letting the joint be hyperextended.

Concept #2—Kinetic Energy: *Kinetic energy* is defined to be equal to the mass of the striking object times the square of the velocity of the object divided by two. In other words, it's important to have body mass behind a strike, which is why tensing static muscles is important—this mechanically connects the body's mass to the blow. For example, if you strike with a fist but don't use the static muscles of the shoulder and torso, then you might generate one unit of energy because only the mass of the fist and the forearm contribute to the blow. If you tense the upper arm and shoulder during impact, the effective mass of the strike could easily go up by a factor of five, as could the amount of energy generated. However, it's even more important to have good velocity behind a strike—if you double the speed of the blow, the amount of energy would go up by a factor of four (two squared). Thus, if you increased the effective mass or body mass by a factor of five and doubled the blow's speed, the amount of the energy in the blow could go up by as much as a factor of twenty (five times two squared).

The bottom line is that it's important to increase both the effective mass and the speed behind a blow. The problem is that to increase the effective mass of the blow, you must tighten the correct static muscles; tightening the wrong muscles will slow down the strike. On the other hand, to increase the strike's velocity, the dynamic muscles must be tensed and the opposing muscles must be relaxed, which will decrease the effective mass of the blow. Thus, when you want to increase a blow's energy, there's an intricate trade-off between the effort to increase the blow's effective mass and the effort to increase the velocity of the striking surface. The timing involved in tensing both the dynamic and static muscles is critical. However, given a choice, increasing speed usually proves more effective in magnifying the energy of a blow.

Concept #3—Elastic versus Inelastic Collisions: A strike has a certain amount of inherent energy. The laws of physics require that the energy goes somewhere since energy is always conserved: It might go from the striking surface into the target and cause damage to the target; it might go from the striking surface into the target and cause the target to fly, undamaged, backward (it may get damaged when it falls to the floor or hits a wall, but that's a different story); or the striking surface may hit a hard, immobile object and the striking surface will either be damaged or perhaps just bounce off the target. How often have you seen a beginner walk up to a swinging heavy bag and give it a good whack, only to find himself flung back and the heavy bag continuing to swing, relatively unimpaired? This is an example of an elastic collision, something martial artists hope to avoid. The following are a couple of traditional, physics-based examples of elastic and inelastic collisions of two rolling balls.

Example #1 (elastic collision): Take two billiard balls and bounce them off each other. They will fly away from each other at the same relative speed at which they struck, and no damage will occur to either ball. **Example #2 (inelastic collision):** Take a billiard ball and a clay ball and roll them toward each other. The two balls will become one mass as the clay ball is distorted by some of the energy of the collision; the rest of the energy propels the resulting mass away at a reduced speed.

Example #1 is what commonly happens with beginner martial artists—their strikes are ineffective. Example #2 is what a martial artist would like to achieve.

Author Lily Chou wins yet another point from fellow author Norman Link.

HOW TO USE THIS BOOK

This book contains illustrations of 50 common martial arts techniques. While there are countless ways of executing many of these techniques, such as a punch, we focus on the basic, universal elements shared by different styles. It's not the purpose of this book to teach the techniques featured here. Rather, we point out key aspects of the techniques that are responsible for speed, power, and accuracy. Although numerous muscles are necessary to perform a technique, we only identify the primary muscles involved. In the illustrations, the red muscles signify key dynamic, or moving, muscles, while the blue muscles are key static muscles, or muscles that are tensed but not moving. Note that the status of the key muscles usually changes as the technique progresses from the beginning to the end.

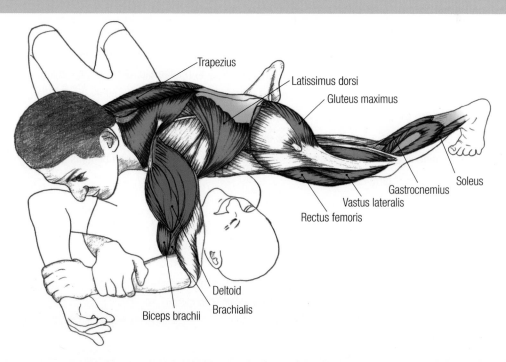

Trapezius

Latissimus dorsi

Gluteus maximus

Soleus

Gastrocnemius

Vastus lateralis

Rectus femoris

Deltoid

Brachialis

Biceps brachii

Each technique features short descriptions of three critical characteristics of the technique: the relative *Speed* of the technique, the relative *Power* required to execute the technique, and the *Accuracy* required to implement the technique. Each of these descriptions is accompanied by a rating such as [2 of 10] or [9 of 10]. These ratings reflect the authors' opinion of the relative importance of each of these areas for an average practitioner. For example, an upper block's power requirement is rated 9 of 10 to emphasize the importance of the technique's power aspect, but the speed rating for the block is only 5 of 10. This isn't to say that speed isn't important, but we feel that it's not nearly as important as the technique's power portion.

Each illustration is paired with several conditioning exercises and stretches that target these key muscles to aid in the development of power and/or speed. Some of these exercises you might encounter solely in a martial arts setting or are a twist on common exercises and yoga poses; you may recognize the bulk of the exercises and stretches from strength-training, yoga, or even grade-school P.E. classes. Since thousands of books and videos specialize in the various realms of fitness (in particular, we suggest Lily Chou's *The Martial Artist's Book of Yoga* for detailed information on the yoga poses, and Bob Anderson's *Stretching*), we provide instructions only for the lesser-known movements (see the Appendix, page 128).

You may choose to integrate these exercises and stretches into your regular exercise routines or make them a separate workout. The number of reps and sets you do is a matter of individual preference, but there are basic rules to follow, depending on whether you're training to improve speed, power, or both. Many repetitions with lighter weights are generally used to build speed, while fewer repetitions with heavier weights are used to build power. Plyometric exercises such as burpies and clapping push-ups enhance both speed and power.

Remember to let your muscles rest for at least 24 hours after a muscle-building session (more if you're doing high-intensity plyometrics)—rest allows your muscles to grow and repair themselves. Regularly rotating the exercises practiced can also give the different muscle groups an opportunity to recuperate. Strapping on ankle or wrist weights is usually acceptable for slow, muscle-building work but not for fast movements; the joints, especially the elbows and knees, can be damaged by numerous hyperextensions. While it's true that the limbs feel light after taking the weights off, this form of practice places a major strain on the elbows and knees and a number of injuries have been attributed to this practice. In general, the use of leg weights should be avoided, as the short-term benefits are outweighed by the possibility of sustaining long-term damage.

! Some of the exercises call for resistance bands, dumbbells, and medicine balls; because resistance bands can snap and cause serious injuries to the eyes and elsewhere, make sure that your bands are in good condition.

PART 2: TECHNIQUES

HAND STRIKES & BLOCKS

Hand strikes and blocks require a precisely orchestrated flow of coordinated power and speed that usually starts from the feet and legs, and is focused up through the body and out the striking hand. Hand techniques are taught with varying proportions of speed and power depending on a student's physical assets.

Hand strikes are generally faster and more accurate than kicks. Since the arm's mass is about half that of the leg, this additional speed makes up for the arm's relative lack of mass. Correct body balance and pivoting are the keys to delivering an effective hand strike. Other major factors include the size of the surface area of the blow (a punch with two knuckles is often more effective than an equivalent blow with a palm heel strike) and the addition of mass (i.e., power) by "putting your body into the blow." These factors are discussed in "The Physics Behind a High-Energy Strike" (page 11) but should be reviewed in detail with your instructor.

While kicks are often stronger, hand techniques—including blocks—can be delivered with a great deal of power. How much force can a "good" strike generate? The simple answer is that a strong hand strike such as a reverse punch rarely exceeds 1000 pounds of force, while a strong kick such as a turning side kick can generate 2000 pounds of force. Brick breaks are often used to demonstrate hand-strike proficiency with an emphasis on both speed and power: A standard palm heel brick break emphasizes more power, while a handstand brick break emphasizes more speed and timing. Timing, as well as power and speed, is critical when techniques are applied to an attacker.

Due to space limitations, this book has not dedicated much space to the critical concepts of breathing and yelling (*kihap*, *kiai*, or spirit yell). This subject cannot be overemphasized, as it helps concentrate and coordinate the body's actions and tenses the core muscles, creating a more rigid base from which to move the extremities. Given that the arms are weaker than the legs, these concepts have added importance for hand strikes and blocks.

Hand Strikes & Blocks

- Front Punch
- Reverse Punch
- Palm Heel
- Lead-Hand Back Knuckle

- Knifehand Chop
- Front Elbow
- Downward Block
- Upper Block

- In-to-Out Block
- Out-to-In Block
- Palm Heel Brick Break
- Handstand Brick Break

FRONT PUNCH

The fastest of the straight punches, the front punch uses speed in lieu of power. Because the arm travels a relatively short distance, it's easy to surprise someone with this technique and thus it's frequently used as an initial move. This attack is usually aimed at the face or used as a setup for a second, stronger technique.

Speed (9 of 10)

Upper-body movement creates most of the speed for this strike. To minimize your opponent's ability to detect that the punch is coming, body movements (such as the hip and shoulder turn) should be subtle; the arm extension is the only big motion involved.

Power (4 of 10)

While most people stress speed over power with this strike and work on the speed of the shoulder twist and arm extension, some schools emphasize a small but powerful twist of the hip to allow the back leg to drive into the blow. Some key factors in power generation are:

Body drive: This move's power relies heavily on the posterior and shoulder kinetic chains working together. While the turn of your shoulder into the blow is subtle, the tensing of the shoulder after the movement is more critical than usual because the driving weight of your upper body is the primary source of power.

Fist pronation: Turning your striking palm downward twists the two bones of the forearm (radius and ulna), making them mechanically much firmer and less elastic. This allows efficient power transfer into the target.

Accuracy (7 of 10)

The relative weakness of the front punch means that accuracy is paramount. Striking an opponent's face is difficult, simply because they can see the blow coming; be careful to not "telegraph" the strike. The angle of your chest to the blow during impact makes a big difference in delivery. At impact, usually the chest turns a little (maybe 30 degrees) from the target. This flatness allows you to be in a good position to execute the maximum number of secondary or follow-up techniques. However, sometimes the chest must be turned much farther (up to 90 degrees), depending on where the target is. This extends the front punch's range but comes at the cost of reducing its power and limiting the number of techniques that can immediately follow. You should practice all chest-angle positions. Working in front of a mirror, using a video recorder, and consulting an instructor are the best ways to refine the front punch while maintaining speed and accuracy.

KEY EXERCISES

Push-up
Strengthens pecs and triceps

Dip
Strengthens triceps

Reverse plank
Stretches arms, shoulders, and front of body

Key Dynamic Muscles

Arm extension: deltoids, triceps, pectorals, serratus anterior

Fist pronation: pronators (unseen)

Body drive: quadriceps, calves

Key Static Muscles

Abdominals, posterior deltoid, gluteus maximus, quadriceps, hamstrings

Primary Kinetic Chains

Posterior, shoulder turn, arm extension

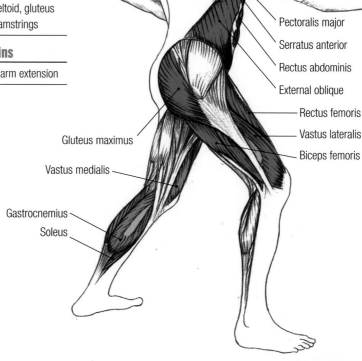

Middle deltoid
Posterior deltoid
Triceps brachii

Pectoralis major
Serratus anterior
Rectus abdominis
External oblique
Rectus femoris
Vastus lateralis
Biceps femoris

Gluteus maximus
Vastus medialis
Gastrocnemius
Soleus

Arm-across-chest stretch
Stretches shoulders

COMMENTS

1) Take care to avoid repeated hyperextension of the elbow during practice. Numerous injuries can result from this abuse, the most common of which is tendinitis. This injury is a bit less common with the reverse punch since the pectorals are more readily available for slowing and controlling the punch.

REVERSE PUNCH

The most powerful of the straight punches, the reverse punch has the ultimate blend of speed and power. Because the striking arm travels a relatively long distance, it's difficult to surprise someone with this technique and thus it's not usually used as an initial move. Common targets range from the face all the way down to the thighs, but the reverse punch is usually aimed at the face or mid-section.

Speed (7 of 10)

The reverse punch is a classic sequence of kinetic-chain movements, and the speed for this strike relies on the interaction among them.

Power (9 of 10)

Some key factors in power generation are:

Body drive: This move's power relies heavily on the posterior, hip turn, and shoulder turn kinetic chains working together.

Arm extension at impact: The two major factors that determine the power of the blow are the velocity of your fist and the mass behind it. In general, it's ideal to impact the target when your fist is moving at maximum speed, which occurs when your arm is about 45 degrees from straight. *Note:* Some schools believe that the blow should impact a bit later; this means your hand may have slowed down a bit, but more of your body's static muscles have had a chance to tense, resulting in more mass behind the blow.

Fist pronation: Turning your striking palm downward twists the two bones of the forearm (radius and ulna), making them mechanically much firmer and less elastic. This allows efficient power transfer into the target.

Accuracy (9 of 10)

While a reverse punch is one of the strongest hand strikes, it may be wasted if it hits a poor target, such as the chest or back. Timing the strike with your opponent's movement is critical: If your opponent moves away at the moment of impact, then the relative velocity and effective mass of the blow is reduced. You can practice timing with a swinging bag or an air shield, but be careful to not collapse or bend your wrist, which can result in a sprain or dislocation.

KEY EXERCISES

Push-up
Strengthens pecs, triceps, and wrist extensors

Dip
Strengthens triceps

Lunge + twist
Enhances hip flexibility while developing core power

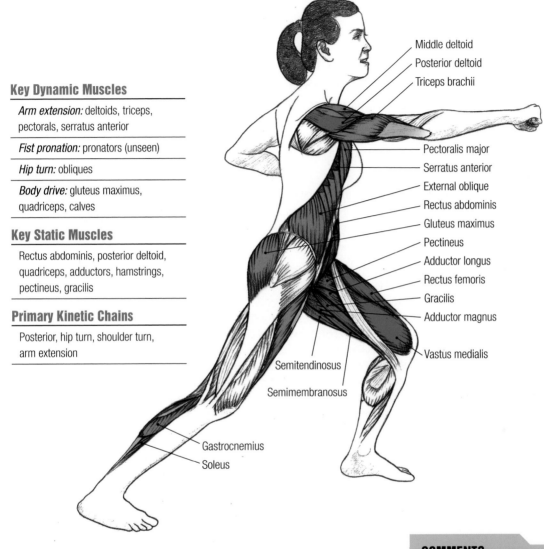

Key Dynamic Muscles

Arm extension: deltoids, triceps, pectorals, serratus anterior

Fist pronation: pronators (unseen)

Hip turn: obliques

Body drive: gluteus maximus, quadriceps, calves

Key Static Muscles

Rectus abdominis, posterior deltoid, quadriceps, adductors, hamstrings, pectineus, gracilis

Primary Kinetic Chains

Posterior, hip turn, shoulder turn, arm extension

Middle deltoid
Posterior deltoid
Triceps brachii
Pectoralis major
Serratus anterior
External oblique
Rectus abdominis
Gluteus maximus
Pectineus
Adductor longus
Rectus femoris
Gracilis
Adductor magnus
Vastus medialis
Semitendinosus
Semimembranosus
Gastrocnemius
Soleus

Warrior 1
Strengthens lower body; stretches quads and shoulders

Reverse plank
Stretches arms, shoulders, and front of body

COMMENTS

1) Because wrist flexors (in the direction of the palm) are almost always stronger than wrist extensors (in the direction of the back of the hand), it's common to hurt your wrist while punching by folding the wrist toward the palm. To guard against this, martial artists who do a lot of punching should occasionally work the wrist extension muscles.

PALM HEEL

This powerful hand strike is taught from both a front and reverse stance. Palm heel from a front stance is faster and less powerful; palm heel from a reverse stance, as illustrated here, is slower but more powerful. Common targets for this attack are the face, chin, solar plexus, and groin.

Speed (9 of 10)

The palm heel can be delivered with the same speed as the reverse punch, with the exception that the hand twist and final position will differ depending on the target. You can increase speed by improving the linearity or directness of the strike and working on pulling your fingers back sharply as the hard butt of your hand strikes the target.

Power (7 of 10)

Locking out the muscles running from your back foot all the way to your striking hand, which maximizes your body weight behind the strike, generates power for this blow. Some key factors in power generation are:

Body drive: The power of this move relies heavily on the posterior, hip turn, and shoulder turn kinetic chains working together.

Arm extension at impact: The two major factors determining the power of the blow are the final velocity of your hand and the effective mass behind the blow. In general, it's optimal to impact the target when your fist is moving at maximum speed, which occurs when your arm is about 45 degrees from straight. *Note:* Some schools believe that the blow should impact a bit later; this means your hand may have slowed down a bit, but more of your body's static muscles have had a chance to tense, resulting in more mass behind the blow.

Palm pronation: Turning your palm down twists the two bones of the forearm (radius and ulna), making them mechanically much firmer and less elastic. This allows efficient power transfer into the target. However, pronating the hand for a palm heel strike is not always possible, depending on the target of the strike.

Accuracy (9 of 10)

This strike is strong, but it will be wasted if it hits a poor target, such as the chest or back. Timing the strike with your opponent's movement is critical: If the opponent moves away at the moment of impact, then the relative velocity and effective mass of the blow is reduced.

KEY EXERCISES

Push-up
Strengthens pecs and triceps

Dip
Strengthens triceps

Lunge + twist
Enhances hip flexibility while developing core power

Key Dynamic Muscles

Arm extension: deltoids, triceps, anconeus, trapezius, serratus anterior

Palm strike: pronators (unseen), wrist extensors

Body drive: gluteus maximus, quadriceps, calves

Key Static Muscles

Abdominals

Primary Kinetic Chains

Posterior, hip turn, shoulder turn, arm extension

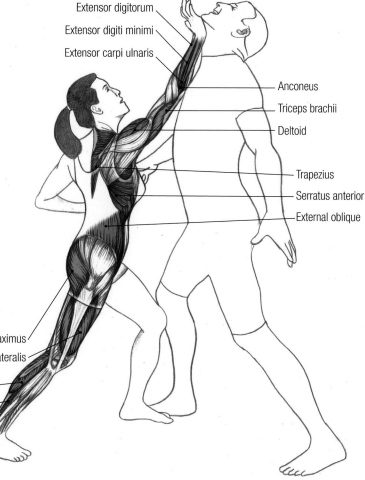

Extensor digitorum

Extensor digiti minimi

Extensor carpi ulnaris

Anconeus

Triceps brachii

Deltoid

Trapezius

Serratus anterior

External oblique

Gluteus maximus

Vastus lateralis

Gastrocnemius

Soleus

Reverse plank
Stretches arms, shoulders, and front of body

Kneeling forearm stretch
Stretches wrists and forearms

COMMENTS

1) The palm heel strike is recommended instead of a fist strike for brick and board strikes because it saves the knuckles from long-term damage.

2) When striking, be careful not to strike on the thumb side of your palm heel, as the nerves to the thumb can be damaged.

3) If your fingers aren't pulled back enough, sometimes the fingertips strike the target before the palm does, which reduces the strike's effectiveness.

LEAD-HAND BACK KNUCKLE

One of the fastest hand strikes, the lead-hand back knuckle has a great deal of speed, plus enough power to stun—if not knock out—an opponent. Because of the short distance traveled and its speed, it's usually used as an initial technique. The target is often the head, but a common variation also attacks the groin. Spinning and turning versions are also taught.

Speed (7 of 10)

The timing of your hand's speed and your body's forward lunge are critical in an effective back knuckle strike. This technique relies heavily on body drive and twist.

Power (6 of 10)

Your fist's speed generates most of the power in the lead-hand back knuckle, as there is little body weight behind the blow. The turning or spinning back knuckle is much more powerful because there's more body mass attached to it. Some key factors in power generation include:

Arm extension: The arm's snapping extension at the shoulder and elbow generates the majority of the blow's power. While this is commonly attributed to the shoulder turn kinetic chain, both the lateral and hip turn kinetic chains are also critical.

Wrist snap: At the moment of impact, the wrist, which is initially flexed, is snapped straight, giving a whiplike power to the final segment of the blow.

Accuracy (9 of 10)

Accuracy is vital since the back knuckle is not very powerful and is thus only effective on a limited number of targets. You can increase your accuracy with drills, such as having a partner hold two hand paddles and flash them out at face and groin height for quick strikes.

KEY EXERCISES

Warrior 2 band pull (page 129)
Strengthens legs, hips, shoulders, and triceps; stretches chest

Lunge + twist
Enhances hip flexibility while developing core power

Wide-leg forward bend + shoulder stretch
Stretches hamstrings, adductors, and shoulders

Key Dynamic Muscles

Arm extension: trapezius, rhomboids, triceps, anconeus, deltoids

Wrist snap: wrist extensors

Body drive: gluteus maximus, gluteus medius, quadriceps (unseen), calves

Body twist: obliques

Key Static Muscles

Posterior deltoid, obliques

Primary Kinetic Chains

Lateral, hip turn, shoulder turn, arm extension

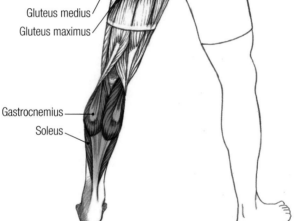

Middle deltoid
Posterior deltoid
Extensor digitorum
Anconeus
Triceps brachii
Trapezius
Rhomboids
External oblique
Gluteus medius
Gluteus maximus
Gastrocnemius
Soleus

Arm-across-chest stretch
Stretches shoulders

COMMENTS

1) Some schools teach the turning or spinning back knuckle with a bottom fist, as this position protects the elbow from hyperextending if the technique is blocked or jammed.

KNIFEHAND CHOP

The tensed knife edge of the hand delivers this fairly fast, hard blow. Since the strike usually doesn't have much power behind it, it's often used for small, specific targets. The illustrated technique shows a strike that's meant to break an opponent's collarbone.

Speed (7 of 10)

Arm extension, shoulder turn, and the final snap of the wrist to the ulnar (little finger) side generate most of the speed in this technique. The wrist snap and tightening of the knife edge of the hand are required for fast, strong delivery.

Power (4 of 10)

In a self-defense situation, as shown here, this strike's power isn't usually quite as much as other hand strikes' because, as a swinging blow, it doesn't have as much body weight behind it. That said, experienced practitioners *can* wind up and swing the arm in a long arc and break numerous boards, bricks, or ice slabs. The difference is that in a self-defense situation, there isn't enough time or space to wind up for such a blow and, even if there was, an opponent would see it coming and block it.

Accuracy (8 of 10)

The relative weakness of the chop means that there's a real premium placed on accuracy. The collarbone is perhaps the most popular target, and many argue about the exact point where the strike should be made in order to break it. However, for most self-defense situations in which the practitioners are not highly trained, we recommend just aiming for the middle of the collarbone. There are numerous alternate targets (such as the temple, corner of the jaw, side of neck, floating ribs, groin, or outer ribs), but most are small and require practice to hit effectively.

KEY EXERCISES

Sit-up with punch (page 129)
Strengthens core and striking muscles; enhances torso flexibility

Lunge + twist
Enhances hip flexibility while developing core power

Wide-leg forward bend + shoulder stretch
Stretches hamstrings, adductors, and shoulders

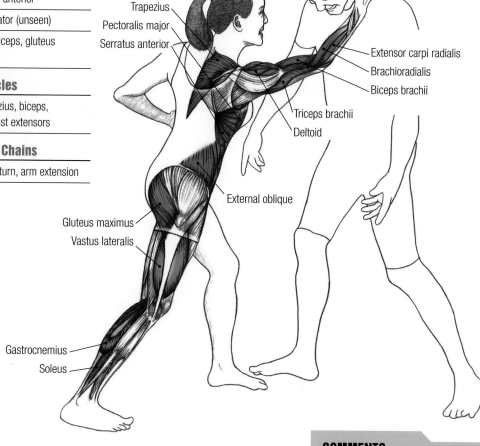

Key Dynamic Muscles

Arm extension: deltoids, triceps, pectorals, serratus anterior

Wrist snap: supinator (unseen)

Body drive: quadriceps, gluteus maximus, calves

Key Static Muscles

Abdominals, trapezius, biceps, brachioradialis, wrist extensors

Primary Kinetic Chains

Hip turn, shoulder turn, arm extension

Trapezius
Pectoralis major
Serratus anterior
Extensor carpi radialis
Brachioradialis
Biceps brachii
Triceps brachii
Deltoid
External oblique
Gluteus maximus
Vastus lateralis
Gastrocnemius
Soleus

High-elbow shoulder stretch
Stretches shoulders and triceps

COMMENTS

1) The chopping nature of this blow requires precise coordination of the shoulder turn, arm extension, and wrist snap. However, unlike most other hand strikes, only the shoulders, torso, hand, and legs are tensed—not necessarily the arm.

2) The intricacies of tensing a knife hand are beyond the scope of this book. However, it's an essential component in delivering a sharp blow to the target.

FRONT ELBOW

Perhaps the most powerful of the hand and arm strikes, the straight front elbow strike is very strong but by its nature is not very fast and is quite short-ranged. It's primarily a close-range self-defense technique.

Speed (3 of 10)

The speed of the elbow strike is relatively slow compared to a hand or foot strike. However, because the blow is used in close quarters, its speed is less important than its power.

Power (9 of 10)

Some key factors in power generation include:

Shoulder turn: While this movement primarily involves the shoulder-turn kinetic chain, numerous variations include pulling the target in with your non-striking hand or using your non-striking hand to pull your striking arm across the target.

Chest angle: To ensure that there's solid body weight behind the blow, your chest should be turned in toward the point of contact.

Arm flexion: Arm flexion is primarily driven by the pectorals and the anterior deltoid, but is also assisted by pulling your hand in sharply toward your chest.

Accuracy (5 of 10)

Because an elbow strike is relatively short-ranged compared to a hand or foot strike, accuracy is important. This short range means it can miss a target or be easily jammed. If correctly placed, it can be very powerful, but an elbow strike is wasted if it hits the flat of an opponent's chest. An elbow strike to most regions of the head should be effective, but since it's slow and an opponent is likely to see it as it comes in, the head can be a difficult target to strike.

KEY EXERCISES

Sit-up with punch (page 129)
Strengthens core and punching muscles; enhances torso flexibility

Lunge + twist
Enhances hip flexibility while developing core power

Warrior 1
Strengthens lower body; stretches quads and shoulders

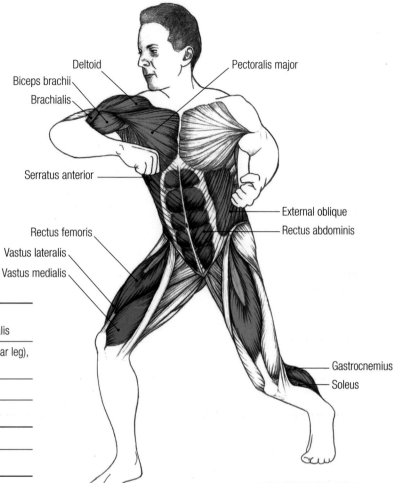

Deltoid

Biceps brachii

Brachialis

Pectoralis major

Serratus anterior

External oblique

Rectus abdominis

Rectus femoris

Vastus lateralis

Vastus medialis

Gastrocnemius

Soleus

Key Dynamic Muscles

Elbow drive: pectorals, serratus anterior, deltoids, biceps, brachialis

Body drive: gluteus maximus (rear leg), quadriceps, calves

Body twist: obliques

Key Static Muscles

Abdominals, quadriceps

Primary Kinetic Chains

Posterior, hip turn, shoulder turn

Wide-leg forward bend + shoulder stretch
Stretches hamstrings, adductors, and shoulders

High-elbow shoulder stretch
Stretches shoulders and triceps

COMMENTS

1) The striking surface of an elbow strike should be at least one inch below the point of the elbow (toward the hand).

2) Striking with the point of the elbow can damage your elbow and must be avoided.

3) Back elbow strikes, where the striking surface is an inch or more *above* the elbow, are harder to master because there are a number of nerves and muscle insertions there that shouldn't be struck.

DOWNWARD BLOCK

This classic, strong block is used primarily against kicking attacks, and there are two major variations: hard and soft. The hard version, as shown here, is a power technique meant to strike to one side of an attacking leg. The soft version is a timing technique used for deflection. Due to the relative power of kicking attacks, smaller people tend to use the soft version of the block more often to avoid risking a broken arm.

Speed (5 of 10)

Arm extension generates most of the block's speed, with additional speed from the shoulder turn. Because the block must intercept an incoming strike, speed and accuracy combine to create an effective movement.

Power (7 of 10)

Power for this block comes primarily from turning the shoulder and dropping weight into the block. Power can also be generated from leg extension and hip turn, but in most actual applications it's difficult to be in position to use these movements. Other key factors in power generation include:

Arm pronation: The pronation, or twist, of the forearm is critical during a hard downward block to effectively drop weight into the block.

Blocking angle: As the block intercepts the kick, the forearm's angle in relation to the kick's line of movement will determine how much the kick will be blocked and how much it will be deflected.

Accuracy (9 of 10)

As with any block, accuracy is essential. Given that this block is usually used against kicks, it's even more important to time it correctly. Many people teach the downward block in such a way that if the timing is off and your opponent's kick connects, your body will twist to the side and not take the full brunt of the blow.

KEY EXERCISES

Cross-body downward band pull (page 128)
Strengthens lats, delts, and triceps

Lunge + twist
Enhances hip flexibility while developing core power

Warrior 1
Strengthens lower body; stretches quads and shoulders

Key Dynamic Muscles

Arm extension: latissumus dorsi (unseen), trapezius (unseen), deltoids, triceps, obliques

Arm pronation: pronators (unseen)

Body drive: calves

Key Static Muscles

Rectus abdominis, gluteus maximus (front leg), rectus femoris

Primary Kinetic Chains

Shoulder turn, arm extension

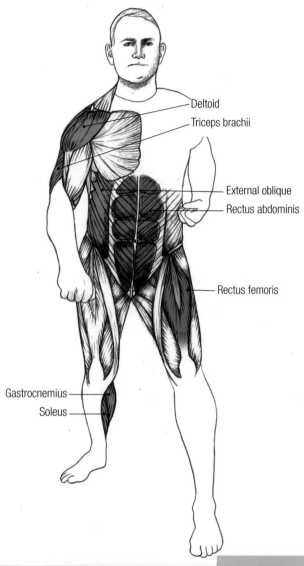

Deltoid

Triceps brachii

External oblique

Rectus abdominis

Rectus femoris

Gastrocnemius

Soleus

Reverse plank
Stretches arms, shoulders, and front of body

High-elbow shoulder stretch
Stretches shoulders and triceps

COMMENTS

1) While this tends to be executed as a hard block, with careful timing it can be turned into a soft block and twisted into a catch-and-trap technique. However, softening the block to allow a catch often results in a poorly executed block, which means the defender gets struck by the attack.

UPPER BLOCK

This classic, strong block is used primarily against downward strikes to the head; it combines power (to slow the blow) and deflection (to ward off the blow). Because any head strike is dangerous, this block is important to learn. As with many other blocks, smaller people tend to use the upper block to deflect the blow and avoid risking a broken arm with a hard block.

Speed (5 of 10)

Arm extension generates most of the block's speed, with additional speed from an early forward drive of the hips. Given that the block must intercept an incoming strike, speed and accuracy combine to create an effective block.

Power (9 of 10)

This block's power is generated primarily from the upward drive of the blocking arm. Power also comes from hip extension and from locking out the entire body from the point of the block all the way down to the rear leg. Other key factors in power generation include:

Arm pronation: The pronation, or twist, of the forearm is critical during a hard block to effectively push your weight into the block.

Blocking angle: As the block intercepts the blow, the angle of the forearm in relation to the blow's line of movement will determine how much the strike will be blocked and how much it will be deflected.

Accuracy (6 of 10)

As with any block, accuracy is essential. Given that this block is usually used against a blow to the head, it's even more important to time it correctly. Many people teach the upper block in such a way that if it's only partially effective and the blow bounces off the blocking arm, the shoulder—not the head—will take the brunt of the blow.

KEY EXERCISES

Barbell/dumbbell pullover
Strengthens pecs, triceps, and lats

Plank
Strengthens core and deltoids

Warrior 1
Strengthens lower body; stretches quads and shoulders

Key Dynamic Muscles

Arm extension: trapezius, deltoids, triceps

Arm pronation: pronators (unseen)

Body drive: quadriceps (unseen)

Key Static Muscles

Latissimus dorsi, serratus anterior, gluteus maximus, hamstrings

Primary kinetic chains

Posterior, arm extension

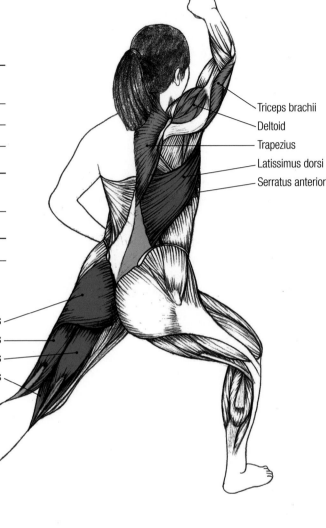

Triceps brachii
Deltoid
Trapezius
Latissimus dorsi
Serratus anterior

Gluteus maximus
Biceps femoris
Semitendinosus
Semimembranosus

Reverse plank
Stretches arms, shoulders, and front of body

High-elbow shoulder stretch
Stretches shoulders and triceps

COMMENTS

1) Unlike some hard blocks (such as a cross block), the upper block will almost never stop a blow. Its purpose is to deflect the blow. Upon finishing the block, you should be prepared to take advantage of your attacker's momentum and balance shift to immediately begin a control or counterstrike.

IN-TO-OUT BLOCK

The weaker of the two classic mid-body blocks (the other being the out-to-in block on page 36), this technique can be performed either as a soft or hard block. Supinating the blocking hand (turning it toward the body, as shown) usually makes the block hard. Pronating the blocking hand (turning it away from the body) allows the block to be either hard or soft.

Speed (5 of 10)

The hip turn, shoulder turn, and external rotation of the shoulder generate most of the speed in this block. While speed isn't as important in this technique, the timing is critical.

Power (5 of 10)

The relative weakness of this block makes it necessary to move your body into the block and then tighten your upper body in order to put as much weight as possible behind the block. Other key factors in power generation include:

Fist supination: Twisting the palm inward (toward your body) tightens the forearm and gives a sharper delivery of power during the block.

Shoulder turn: While the hip turn and the arm's outward rotation are important, the weight derived from tensing your shoulders is the main power source.

Accuracy (6 of 10)

This block's weakness makes it necessary to block farther from your body so that there's more time for the deflected strike to move past you. This means the block must be stretched out, away from the body, thus making it even weaker. This trade-off is difficult to master.

KEY EXERCISES

In-to-out band pull (page 128)
Strengthens traps, rhomboids, and delts

Lunge + twist
Enhances hip flexibility while developing core power

Wide-leg forward bend + shoulder stretch
Stretches hamstrings, adductors, and shoulders

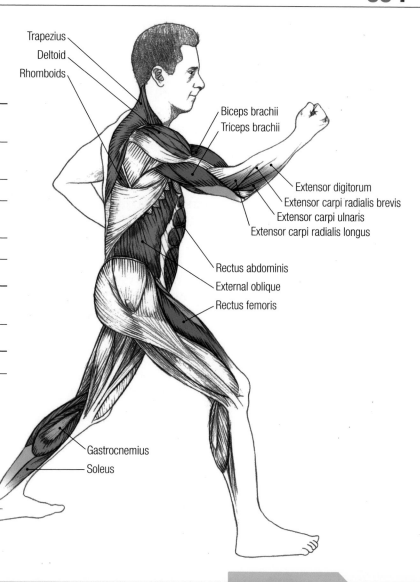

Trapezius
Deltoid
Rhomboids
Biceps brachii
Triceps brachii
Extensor digitorum
Extensor carpi radialis brevis
Extensor carpi ulnaris
Extensor carpi radialis longus
Rectus abdominis
External oblique
Rectus femoris
Gastrocnemius
Soleus

Key Dynamic Muscles

Fist supination: supinator (unseen), biceps

Shoulder external rotation: trapezius, rhomboids, deltoids

Shoulder turn: obliques

Body extension: gluteus maximus (rear leg)

Body drive: calves

Key Static Muscles

Rectus abdominis, triceps, wrist extensors, rectus femoris

Primary Kinetic Chains

Posterior, hip turn, shoulder turn

High-elbow shoulder stretch
Stretches shoulders and triceps

COMMENTS

1) The most common form of a soft in-to-out block leads directly to grabbing the opponent and immediately counterstriking with a reverse punch, knee lift, etc.

2) Since the biceps brachii wraps around the radius, this muscle is not only a powerful elbow flexor but also a very strong supinator.

OUT-TO-IN BLOCK

The out-to-in block is the stronger of the two classic mid-body blocks (the other being the in-to-out block on page 34), but successful timing in intercepting an opponent's blow makes it difficult to master.

Speed (5 of 10)

The hip turn, shoulder turn, and internal shoulder rotation generate most of the block's speed. While speed isn't key in this technique, the timing is critical.

Power (6 of 10)

Timing this block is difficult, making it necessary to move your body into the block and then to tighten your upper body in order to put as much weight as possible behind the block. Other key factors in power generation include:

Fist supination: Twisting the palm inward (toward your body) tightens the forearm and gives a sharper delivery of power during the block.

Shoulder twist: While the hip turn and the arm's inward rotation are important, the weight derived from tensing your shoulders is the main power source.

Accuracy (6 of 10)

The relative weakness of this block makes it necessary to block farther from your body so there's more time for the deflected strike to move past you. This means the block must be stretched out, away from the body, thus making it even weaker. This trade-off is difficult to master.

KEY EXERCISES

Dumbbell fly
Strengthens pecs

Sit-up with punch (page 129)
Strengthens core and striking muscles; enhances torso flexibility

Lunge + twist
Enhances hip flexibility while developing core power

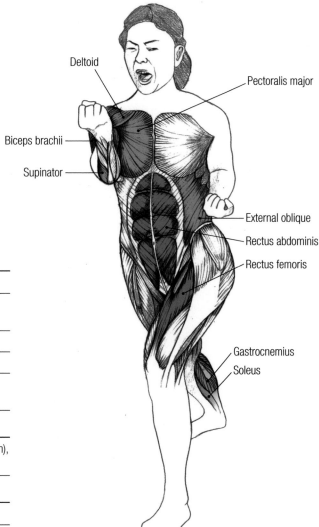

Deltoid

Pectoralis major

Biceps brachii

Supinator

External oblique

Rectus abdominis

Rectus femoris

Gastrocnemius

Soleus

Key Dynamic Muscles

Fist supination: supinator, biceps

Shoulder internal rotation: pectorals, deltoids

Shoulder turn: obliques

Body drive: calves

Body extension: gluteus maximus (rear leg)

Key Static Muscles

Rectus abdominis, teres minor (unseen), rectus femoris

Primary Kinetic Chains

Posterior, hip turn, shoulder turn

Wide-leg forward bend + shoulder stretch
Stretches hamstrings, adductors, and shoulders

Arm-across-chest stretch
Stretches shoulders

COMMENTS

1) In this block, your weight is usually forward, so it's possible to pivot on your front leg if the blow passes the block. The partially blocked blow will then hit you off center and its energy will spin you, which lessens the blow's impact.

PALM HEEL BRICK BREAK

This demonstration technique shows concentration and focus of energy.

Speed (6 of 10)

While hand speed is critical, it needs to be combined with proper use and alignment of body weight for a successful break, especially when breaking larger numbers of bricks. Hand speed doesn't have to be exceptionally fast, but it does have to be timed with the drop of the body weight to strike the bricks at the proper time.

Power (8 of 10)

Some key factors in power generation include:

Arm extension: People strike in at least two distinct ways. Some power their arms through the stack; some snap their arms out and back, as with a jab. The snapping motion should only be used for speed breaks.

Body drop: Both legs must bend simultaneously so your body can drop evenly. People commonly straighten one or both legs and thus move the body mass away from the centerline of the break; this weakens the strike because the weight is no longer moving directly through the line of the break.

Shoulder turn: The turn of the shoulders in conjunction with the dropping of the body is critical in developing maximum power.

Accuracy (7 of 10)

Striking the brick on the centerline, one-third of the distance from the front edge, is considered optimal for breaking a brick. In the final preparation, the elbow of the striking hand must be directly over the point of contact on the brick. When breaking more than one brick, the power line must pass through all the bricks to ensure that they're all broken. Often, when the elbow drifts back on the final upswing, only the top bricks are broken, and it's not uncommon for the palm side of the forearm to scrape across the lower bricks.

KEY EXERCISES

One-arm dumbbell row
Strengthens trapezius

Palm heel strike to targets
Increases arm speed and strength

Push-up
Strengthens pecs and triceps

Key Dynamic Muscles

Arm extension: pectorals, trapezius, deltoids, triceps, wrist extensors

Arm pronation: pronators

Key Static Muscles

Rectus abdominis, trapezius, quadriceps (front leg)

Primary Kinetic Chains

Hip turn, shoulder turn, arm extension

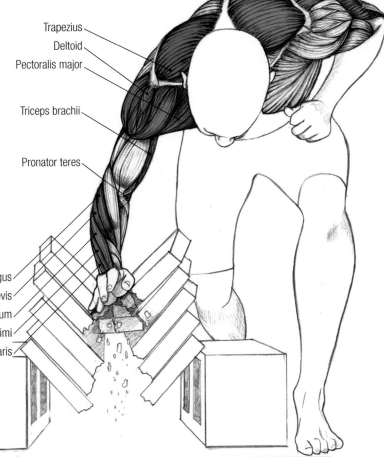

Trapezius

Deltoid

Pectoralis major

Triceps brachii

Pronator teres

Extensor carpi radialis longus

Extensor carpi radialis brevis

Extensor digitorum

Extensor carpi digiti minimi

Extensor carpi ulnaris

Lunge + twist
Enhances hip flexibility while developing core power

Cobra
Stretches chest, shoulders, and abs

COMMENTS

1) Striking with the palm heel on the thumb side is very dangerous—it can cause damage to the nerves in the thumb. This can lead directly to the weakening or loss of use of the thumb.

2) While some people prefer to strike with a punch, the risk of long-term knuckle damage makes the palm heel the preferred technique.

3) Identical-looking bricks can break with vastly different forces. Before breaking stacks of bricks, break just one to test the materials and avoid injuries.

HANDSTAND BRICK BREAK

This demonstration technique requires unusual balance, timing, concentration, and focus of energy. Unlike a standard brick break performed with both feet on the ground, this break involves kicking into a handstand and then executing a fast, hard strike while rolling forward and out of harm's way. A very dangerous break, it demands careful preparation to practice safely.

Speed (6 of 10)

While hand speed is critical, it needs to be combined with proper use and alignment of body weight for a successful break. The hand doesn't have to be exceptionally fast, but it must be timed with the drop of the body weight to strike the bricks at the right moment.

Power (6 of 10)

Some key factors in power generation include:

Body drop: After kicking into the handstand, the striking hand is lifted high, and the body starts to fall to that side; the palm heel snaps out as the body tenses above it to add power to the blow.

Arm extension: The palm heel strike must be snapped out hard and fast, and then used to roll forward and away from the broken bricks.

Accuracy (8 of 10)

Striking the top brick on the centerline, one-third of the distance from the front edge, is considered optimal for breaking a brick; in the final preparation, the elbow of the striking hand must be directly over the point of contact on the brick. When breaking more than one brick, the power line must pass through all the bricks to ensure that they're all broken. Often, when the elbow drifts back on the upswing, only the top bricks are broken, and it's not uncommon for the palm side of the forearm to scrape across the lower bricks.

KEY EXERCISES

Dip
Strengthens triceps

T push-up (page 129)
Improves core and upper-body strength

Handstand push-up (page 128)
Enhances balance and core and upper-body strength

Key Dynamic Muscles

Arm chamber (not pictured): trapezius, deltoids, latissimus dorsi, biceps

Arm extension: pectorals (unseen), trapezius, deltoids, triceps, anconeus

Arm pronation: pronators (unseen)

Key Static Muscles

Rectus abdominis (unseen), gluteus maximus, latissimus dorsi, rhomboids, teres major, trapezius, deltoids, triceps, biceps, brachioradialis, wrist extensors

Primary Kinetic Chains

Shoulder turn, arm extension

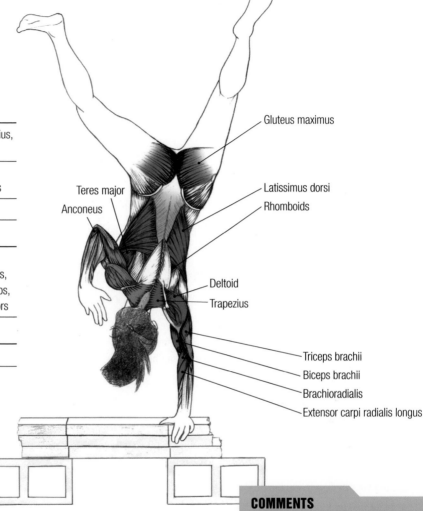

Gluteus maximus

Teres major

Anconeus

Latissimus dorsi

Rhomboids

Deltoid

Trapezius

Triceps brachii

Biceps brachii

Brachioradialis

Extensor carpi radialis longus

Military press
Strengthens delts, pecs, and triceps

High-elbow shoulder stretch
Stretches shoulders and triceps

COMMENTS

1) When kicking into the handstand, you must maintain some forward momentum so that after striking, you can roll forward and away from the bricks. If you kick into a handstand, hold the position, and then strike, there's a tendency to collapse directly down on the bricks, which can lead to great bodily harm.

2) We highly recommend wearing headgear when practicing this technique until you master it.

KICKS

Kicks come in a wide array of styles and variations, including differences in speed, power, and delivery times. Some kicks, such as the front snap, low scoop, and knee lift, are oriented toward self-defense. Others, like the axe, spinning heel, and jumping roundhouse, leave the practitioner open to counterstrikes and thus are used more in arts where the obvious counterstrikes are illegal. It's rare that a single art would find reason to teach all of the techniques featured here. We will offer a small sampling from the broad spectrum of kicks.

Kicks are generally stronger but a bit slower than hand strikes. The mass of the leg is about twice that of the arm; however, the small reduction in speed doesn't overcome the legs' greater mass. As a rule of thumb, kicks are at least twice as powerful as equivalent hand strikes.

Correct balance and pivoting are keys to delivering effective kicks. Timing, power, and speed are critical when kicking an attacker. Board breaks are often used to demonstrate kick proficiency (brick breaks are used less often due to the common occurrence of injuries). The addition of mass (or less exactly "power") by "putting your body into the blow" is a common concern when teaching kicks since a lack of weight behind the blow makes the kick anemic. This and other issues are highlighted briefly in "The Physics Behind a High-Energy Strike" (page 11) but should be discussed in detail with your instructor.

Kicks

- Knee Lift Kick
- Low Scoop Kick
- Front Snap Kick
- Front Thrust Kick
- Roundhouse Kick

- Axe Kick
- In-to-Out Crescent
- Out-to-In Crescent
- Side Kick
- Back Kick

- Jumping Roundhouse Kick
- Spinning Heel Kick
- Low Spinning Heel Kick

KNEE LIFT KICK

This powerful, short-stroke kick relies more on power than on speed or accuracy, and is commonly taught in self-defense classes. The target of this attack ranges from the face all the way down to the thighs, but usually is aimed at the groin or mid-section.

Speed (5 of 10)
The combination of the speed of the hands pulling the opponent down, the acceleration of the knee lift, and the proper use and alignment of the driving knee is required to strike an opponent successfully.

Power (9 of 10)
Some key factors in power generation are:

Hip flexion: The longer the upward swing of the knee, the more powerful the strike will be.

Body drive: As you raise your knee, you must drive yourself toward the target by extending the hip of your supporting leg. Distance from the target is the single most important factor in determining if this kick will be successful. If the opponent is too close, the kick will be jammed; if the opponent is too far away, the kick will miss the target.

Arm and shoulder pull: Pulling the target into the kick adds speed to the strike and is particularly important when kicking a larger opponent.

Accuracy (6 of 10)
Your technique in this kick is more often important than where the blow is positioned on the opponent.

KEY EXERCISES

Mountain climber
Improves lower-body power

Bicycle crunch
Improves core flexibility and strength

Knee raise
(page 129)
Strengthens hip flexors and calves

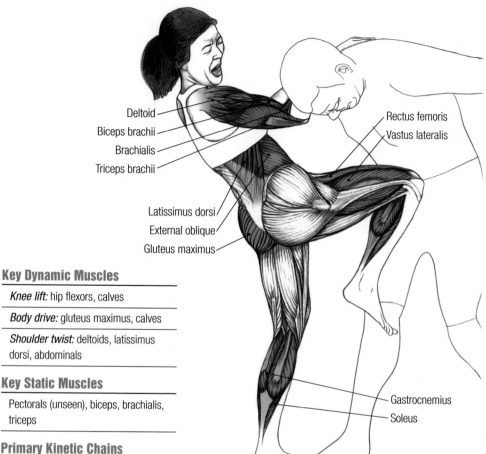

Deltoid

Biceps brachii

Brachialis

Triceps brachii

Rectus femoris

Vastus lateralis

Latissimus dorsi

External oblique

Gluteus maximus

Gastrocnemius

Soleus

Key Dynamic Muscles

Knee lift: hip flexors, calves

Body drive: gluteus maximus, calves

Shoulder twist: deltoids, latissimus dorsi, abdominals

Key Static Muscles

Pectorals (unseen), biceps, brachialis, triceps

Primary Kinetic Chains

Posterior, leg extension (of standing leg), hip turn, shoulder turn

COMMENTS

1) The knee should strike at about the time when the hip is at a 90-degree angle to your body. This is roughly when the knee is moving at its maximum velocity.

2) To generate as much speed and power as possible, the oblique muscles and hip flexors are used to pull the kicking knee up and across your body.

3) Pointing the foot of the kicking leg toward the floor relaxes the hamstring muscles, allowing maximum speed of the rising knee.

Woodchopper (page 129) Strengthens obliques and shoulders

Kneeling lunge Stretches hip flexors and quads

LOW SCOOP KICK

This kick is slow, strong, short-ranged, and self-defense-oriented; it's used primarily against the legs and groin. This is an odd kick in that the striking surface can be the toes, ball of the foot, knife edge of the foot, or heel, depending on the target.

Speed (3 of 10)

The odd in-to-out twist of the leg and hips makes this kick slower than most. While it's usually taught as a type of stomp kick, a close variation snaps the leg out, usually at the groin, like a twisting snap kick. This latter variation is a bit faster. Due to the close range of this kick, speed is not usually a critical factor.

Power (7 of 10)

The kick's stomping action gets much of its power from the stiffening of the entire upper body so that your body weight drives into the blow. Even with all of this weight on the kick, it's usually necessary to strike a leg that has weight on it so that it can't move and will thus take the entire brunt of the blow.

Accuracy (8 of 10)

The accuracy of the blow is found in the importance of delivering a short, sharp kick to an opponent's leg, preferably when the opponent has weight on it. You can enhance the probability of getting weight on the leg if you grab the opponent and pull him forward at the same time as you execute the kick.

KEY EXERCISES

Squat with partner (page 129)
Strengthens quads and glutes

Butterfly
Stretches adductors

Pigeon
Stretches hips, quads, and groin

Key Dynamic Muscles

Leg cocking (not pictured): hamstrings, sartorius

Leg extension: quadriceps, gluteus medius (unseen)

Body twist: obliques (unseen)

Key Static Muscles

Rectus abdominis, quadriceps, hamstrings, calves, gluteus maximus

Primary Kinetic Chains

Hip turn, shoulder turn, leg extension

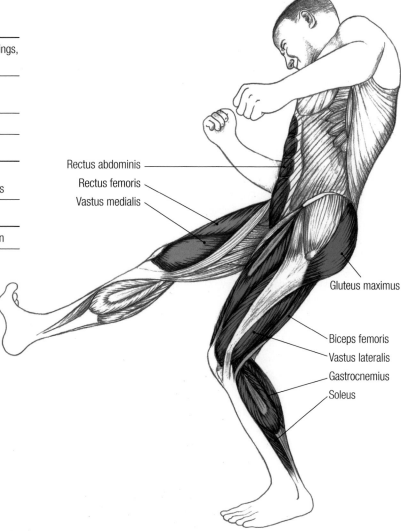

Rectus abdominis

Rectus femoris

Vastus medialis

Gluteus maximus

Biceps femoris

Vastus lateralis

Gastrocnemius

Soleus

COMMENTS

1) While the goal of the kick is to disable your opponent, it's far more likely that you'll merely break his balance. Thus, a follow-up technique such as a throw may be required.

2) A common variation of this kick is performed as a ground defense when you're either on your back or balanced on the non-kicking leg and the opposite hand.

FRONT SNAP KICK

The front kicks, snap and thrust (page 50), are two of the most fundamental kicks in martial arts and thus receive a great deal of attention. The front snap is the faster but less powerful of the two. As such, it's usually thrown with the front leg so maximum speed can be used to get the kick to the target as quickly as possible.

Speed (7 of 10)

Speed is essential in helping to increase the power of the blow. Since this is usually a front-leg kick, the speed of delivery is faster than that of other kicks; for this reason, it's usually thought of as a speed (as opposed to power) kick.

Power (8 of 10)

Power is the direct result of three kinetic chains working in series: The posterior kinetic chain drives the hips forward; the hip turn kinetic chain rotates the hips so that the kicking hip drives forward into the blow; and the leg extension kinetic chain ultimately drives the foot into the target. By tensing the muscles of your torso, primarily the rectus abdominis, the full weight of your upper body helps maximize the blow's power transfer. Note that all three of these chains move less with a front snap kick than with a front thrust kick, thus accounting for less power in this front snap.

Accuracy (6 of 10)

The target of a front snap kick is often somewhere on the centerline of the body such as the groin, stomach, or jaw. While this kick is also effective against other targets, its relative weakness limits the viable options.

KEY EXERCISES

Knee raise (page 129) Strengthens hip flexors and calves

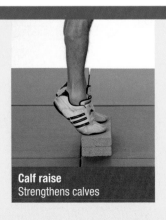

Calf raise Strengthens calves

One-legged bridge + hip dip (page 129) Strengthens pelvic thrust; stretches chest and shoulders

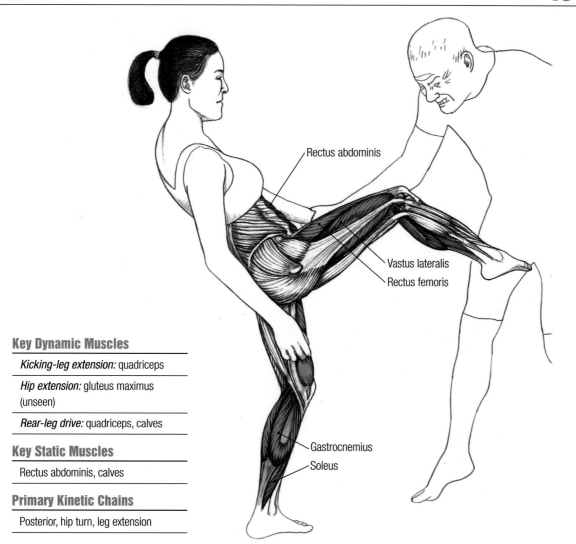

Rectus abdominis

Vastus lateralis

Rectus femoris

Gastrocnemius

Soleus

Key Dynamic Muscles

Kicking-leg extension: quadriceps

Hip extension: gluteus maximus (unseen)

Rear-leg drive: quadriceps, calves

Key Static Muscles

Rectus abdominis, calves

Primary Kinetic Chains

Posterior, hip turn, leg extension

Forward bend
Stretches hamstrings, calves, and hips

Kneeling lunge
Stretches hip flexors and quads

COMMENTS

1) The striking surface on the foot is usually either the instep (used for speed) or the ball of the foot (used for power), depending on the target.

2) When attacking the groin, it's important not to give away your intentions by moving your upper body. Practice fast snap kicks into a mirror and minimize any upper body movement.

FRONT THRUST KICK

The front kicks, snap (page 48) and thrust, are two of the most fundamental kicks in martial arts and thus receive a great deal of attention. The front thrust is the slower but more powerful of the two. As such, it's usually thrown with the back leg so that maximum body-weight shift can be used to increase the kick's power.

Speed (5 of 10)

Speed is very important in helping to increase the power of the blow, but since this is usually executed with the back leg, the speed of delivery is mid-range when compared with other kicks; for this reason, it's usually thought of as a power (as opposed to speed) kick.

Power (8 of 10)

Power is the direct result of three kinetic chains working in series: The posterior kinetic chain drives the hips forward; the hip turn kinetic chain rotates the hips so that the kicking hip drives forward into the blow; and the leg extension kinetic chain ultimately drives the foot into the target. By tensing the muscles of the torso, primarily the rectus abdominis, the full weight of your upper body helps maximize the blow's power transfer.

Accuracy (6 of 10)

The target of a front thrust kick is often the centerline of the body, anywhere from the top of the hip up to the jaw. While the kick is also effective against other targets, such as the kidneys and legs, these are more easily moved out of harm's way and are thus much harder to hit.

KEY EXERCISES

Toe walk (page 129)
Strengthens calves

Burpie
Improves full-body explosive power

Box jump
Improves lower-body explosive power

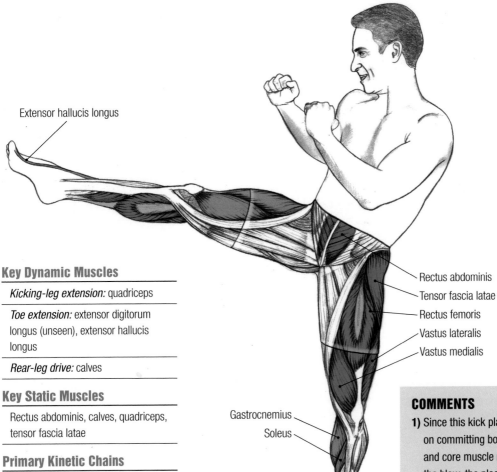

Extensor hallucis longus

Rectus abdominis
Tensor fascia latae
Rectus femoris
Vastus lateralis
Vastus medialis

Gastrocnemius
Soleus

Key Dynamic Muscles

Kicking-leg extension: quadriceps

Toe extension: extensor digitorum longus (unseen), extensor hallucis longus

Rear-leg drive: calves

Key Static Muscles

Rectus abdominis, calves, quadriceps, tensor fascia latae

Primary Kinetic Chains

Posterior, hip turn, leg extension

COMMENTS

1) Since this kick places a premium on committing body weight and core muscle movement to the blow, the placement and orientation of the supporting foot is important. Different styles have the foot ranging from flat on the floor to up on the toes, and pointed anywhere from forward to 135 degrees backward.

2) As with many kicks, the front thrust kick is taught with many striking-surface variations. The most common are the ball of the foot (shown), heel, and knife edge. In self-defense situations, especially with lesser-trained individuals, kicking with the heel is usually safer as there's less chance of ankle injury.

Warrior 1
Strengthens lower body; stretches quads and shoulders

Forward bend
Stretches hamstrings, calves, and hips

ROUNDHOUSE KICK

One of the three "classic" martial arts kicks (the other two being the front thrust kick, on page 50, and side kick, on page 60), the roundhouse has sometimes been described as the leg's equivalent of a slap. It can be performed a variety of ways, including with the front leg (faster) or back leg (stronger). Targets range from the calf to the head.

Speed (9 of 10)

The speed of the roundhouse kick is ultimately a function of the turn of the hips and snapping out of the foot. However, many other actions assist in speed, such as the countertorque generated by the twist of the shoulders and arms.

Power (6 of 10)

Power generation for this kick is difficult to explain because it uses five different kinetic chains. The combination of rapidly turning and then stiffening body parts, ranging from the supporting leg to the hips, torso, and finally leg extension, is not the whole story. You can make a good argument that the movement of the arms and twist of the shoulders is also critical in generating a sharp, powerful kick.

Accuracy (6 of 10)

The coordination of five kinetic chains is an indicator of the complex interaction of the body in this kick. At first blush, one could say that the majority of the kick requires hip and leg coordination. However, so many other body parts move that you could say that virtually all the main parts of the body contribute to the kick's execution. A smooth, powerful roundhouse kick is a common goal of many martial artists, and it takes an unusually long time to develop and maintain.

KEY EXERCISES

Side crunch
Strengthens obliques

Band leg abduction
Strengthens abductors

Warrior 2
Strengthens legs, hips, and shoulders; stretches adductors

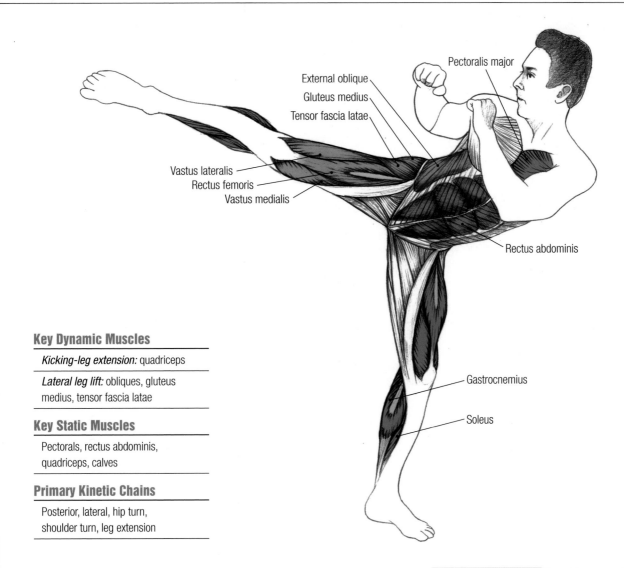

External oblique
Gluteus medius
Tensor fascia latae
Pectoralis major

Vastus lateralis
Rectus femoris
Vastus medialis

Rectus abdominis

Gastrocnemius

Soleus

Key Dynamic Muscles

Kicking-leg extension: quadriceps

Lateral leg lift: obliques, gluteus medius, tensor fascia latae

Key Static Muscles

Pectorals, rectus abdominis, quadriceps, calves

Primary Kinetic Chains

Posterior, lateral, hip turn, shoulder turn, leg extension

Triangle
Strengthens quads; stretches legs, hips, shoulders, chest, and spine

Butterfly
Stretches adductors

COMMENTS

1) The two basic striking surfaces are the top of the foot (for faster kicks) and the ball of the foot (for more damaging impact). Arguments have been made concerning the superiority of each surface, but each has its best uses.

AXE KICK

Usually thought of as a competition or demonstration kick, this technique can be very strong but also makes the kicker vulnerable to counterattacks to the exposed groin and inner leg. Usual targets are the head, collarbone, and, to a lesser extent, the chest.

Speed (5 of 10)

Speed at impact is somewhat dependent on the relative height of the kicker to the opponent. If you're substantially taller than your opponent, the kick often falls from a greater height, has a greater reach, and thus has more time to accelerate. Shorter kickers need to chop down with their leg muscles to compensate for the kick's shorter period of acceleration.

Power (6 of 10)

Leg strike power is generated by pulling your leg down on your opponent and stiffening your leg and body so maximum body weight is behind the kick. Some competition applications require pointing your toes to increase reach, but this substantially reduces the blow's efficacy as it greatly increases the surface area of the impact.

Accuracy (6 of 10)

Striking your opponent's head with this quick kick is difficult to learn because it requires both kicking skill and the ability to time the technique to your opponent's movement. The kick's long, arcing line makes this timing more difficult than with most other kicks.

KEY EXERCISES

Burpie
Improves full-body explosive power

Leg swing forward
Strengthens quads; stretches hips and hamstrings

Hamstring stretch with band
Stretches hamstrings; improves balance

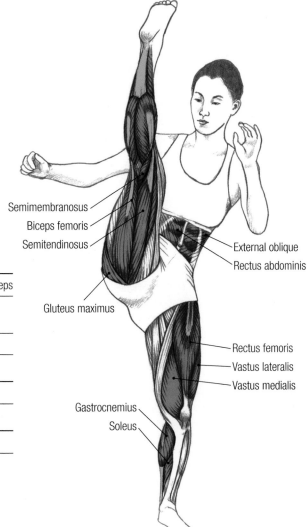

Semimembranosus
Biceps femoris
Semitendinosus
External oblique
Rectus abdominis
Gluteus maximus
Rectus femoris
Vastus lateralis
Vastus medialis
Gastrocnemius
Soleus

Key Dynamic Muscles

Leg chamber (not pictured): quadriceps

Leg strike: gluteus maximus, hamstrings, rectus abdominis

Body drive: quadriceps, calves

Key Static Muscles

Obliques

Primary Kinetic Chains

Posterior, leg extension

Forward bend
Stretches hamstrings, calves, and hips

Kneeling lunge
Stretches hip flexors and quads

COMMENTS

1) An axe kicker's vulnerability to counterattacks can be reduced by adding an initial attack, such as a roundhouse kick or hand strike, to drive the opponent backward.

IN-TO-OUT CRESCENT

The in-to-out and out-to-in crescent (page 58) kicks are often taught together but are of different strengths and tend to have different uses. The in-to-out crescent kick is the stronger of the two because swinging the leg from the inside of the body outward uses the leg abductors, which are significantly stronger than the adductors. This kick is often used for sweeping blocks, whiplike slapping strikes, and as the entrance technique for performing an axe kick (page 54).

Speed (8 of 10)

The hip and shoulder twists and the final outward snap of the kicking leg at the knee generate most of this kick's power.

Power (5 of 10)

This kick's power transfer depends greatly on the part of the kicking foot used. While most people throw the kick such that the knife (outside) edge of the foot strikes the target, this is a large, relatively soft surface. Some people try to strike with the toes turned slightly inward so that the hard edge of the heel strikes the target, increasing the blow's impact. Also note that you can increase power by jumping into or spinning the kick.

Accuracy (3 of 10)

Accuracy isn't hugely important because the sweeping nature of the kick covers a broad area. Some argue that there are exceptions to this, such as during demonstrations, when an in-to-out crescent is used to deflect a hand holding a weapon. Since this isn't a universally accepted use of the kick, we leave this question to the instructor.

KEY EXERCISES

Mountain climber
Improves lower-body power

Leg swing forward
Strengthens quads; stretches hips and hamstrings

Hamstring stretch with band
Stretches hamstrings; improves balance

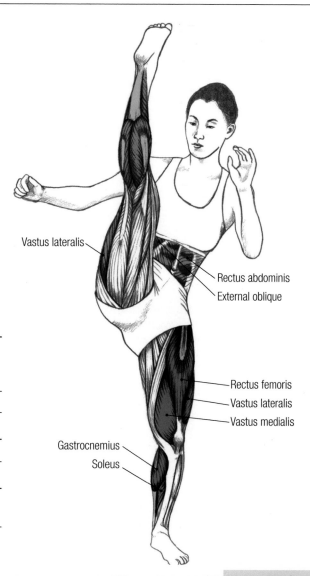

Vastus lateralis

Rectus abdominis

External oblique

Rectus femoris

Vastus lateralis

Vastus medialis

Gastrocnemius

Soleus

Key Dynamic Muscles

Leg extension, foot plantarflexion, and sweep: gluteus medius (unseen), quadriceps, calves

Body twist and rotation: abdominals

Key Static Muscles

Quadriceps, calves

Primary Kinetic Chains

Posterior, hip turn, shoulder turn, leg extension

Wide-leg forward bend
Stretches hamstrings and adductors

Kneeling lunge
Stretches hip flexors and quads

COMMENTS

1) Be careful to avoid having this kick hard-blocked at or near your knee, as severe knee strain can occur.

OUT-TO-IN CRESCENT

The out-to-in and in-to-out crescent (page 56) kicks are often taught together but are of different strengths and tend to have different uses. The out-to-in crescent kick is the weaker of the two because swinging the leg from the outside of the body inward uses the leg adductors, which are not as strong as the abductors (namely the gluteus medius). This kick is often used for sweeping blocks, whiplike slapping strikes, and as the entrance technique for performing an axe kick (page 54).

Speed (7 of 10)

The hip and shoulder twists and the final inward snap of the kicking leg at the knee generate most of this kick's speed.

Power (5 of 10)

This kick's power transfer depends greatly on the part of the kicking foot used. While most people throw the kick such that the bottom edge of the foot strikes the target, this is a large, relatively soft surface. Some people try to strike with the toes turned slightly inward so that the ball of the foot strikes the target, increasing the blow's impact. You can also increase power by jumping into or spinning the kick.

Accuracy (5 of 10)

Accuracy isn't hugely important because the sweeping nature of the kick covers a broad area. There are exceptions to this, such as during demonstrations when an out-to-in crescent is used to deflect a hand holding a weapon. Since this isn't a universally accepted use of the kick, we leave this question to the instructor.

KEY EXERCISES

Mountain climber
Improves lower-body power

Leg swing forward
Strengthens quads; stretches hips and hamstrings

Hamstring stretch with band
Stretches hamstrings; improves balance

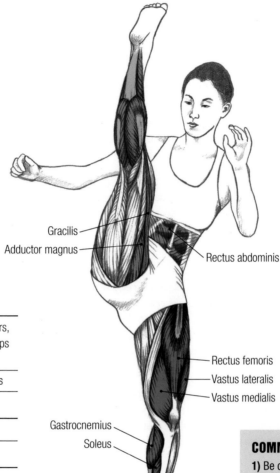

Gracilis

Adductor magnus

Rectus abdominis

Rectus femoris

Vastus lateralis

Vastus medialis

Gastrocnemius

Soleus

Key Dynamic Muscles

Leg extension and sweep: adductors, gracilis, sartorius (unseen), quadriceps (unseen), calves

Body twist and rotation: abdominals

Key Static Muscles

Quadriceps, calves

Primary Kinetic Chains

Posterior, hip turn, shoulder turn, leg extension

Forward bend
Stretches hamstrings, calves, and hips

Kneeling lunge
Stretches hip flexors and quads

COMMENTS

1) Be careful to avoid having this kick hard-blocked at or near the knee as severe knee strain can occur.

2) The obliques on the kicking side initiate the lifting and turning of the hips when starting the kick.

3) Out-to-in crescent is sometimes used in flashy combinations, such as doing an out-to-in crescent that spirals inward and then pulls in and is fired out as a side kick or back kick. This combination of circular and linear kicks requires a complex array of kinetic chains to stop one motion and use the momentum to power into the next movement.

SIDE KICK

This popular kick is one of the three classic martial arts kicks—front (page 50), side, and back (page 62)—that is widely taught and has an incredible number of variations. The side kick combines the accuracy of a front kick with the power of a back kick to make a very accurate, strong technique.

Speed (6 of 10)

In this kick you trade speed for power. Chambering the kick back toward your body before throwing it adds power to the kick but slows its delivery. The front-leg side kick doesn't usually chamber as deeply and is thus very fast but less powerful than the rear-leg version; it's used against the legs and ribs and as a jamming kick.

Power (8 of 10)

The majority of this kick's power is derived from the hip turn and the extension and drive of the kicking leg. Other important contributors are the drive of the supporting leg and the tensing of the torso muscles so that body weight can be added to the kick. The two-step side kick and the turning side kick are two common variations that are significantly more powerful than a standard standing kick. The flying side kick, which requires leaping into the air, is also stronger but has the disadvantage that, once you're in the air, the trajectory of the kick is set and the technique is easier to counter. Higher kicks, as shown, are common in competition but are vulnerable to low counterkicks in self-defense situations.

Accuracy (6 of 10)

Accuracy is important, but this kick's high degree of power makes targeting slightly less important. The coordination of launching the entire body into the kick is critical. Many people get their kicking leg and hip to work well, but they work less hard on coordinating the supporting leg and body weight. For example, if your supporting leg is planted too close to the target, then upon impact you'll be knocked backward because your supporting leg isn't in position to drive into the target.

KEY EXERCISES

Half moon + crunch (page 128)
Strengthens legs, glutes, core, and obliques; stretches legs; improves balance

Side kick extension along wall (page 129)
Strengthens quads, glutes, and obliques

Triangle
Strengthens quads; stretches legs, hips, shoulders, chest, and spine

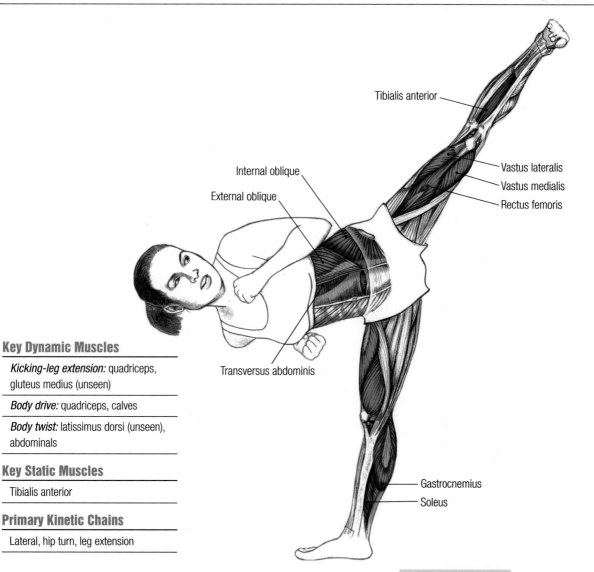

Tibialis anterior

Internal oblique

External oblique

Vastus lateralis

Vastus medialis

Rectus femoris

Transversus abdominis

Gastrocnemius

Soleus

Key Dynamic Muscles

Kicking-leg extension: quadriceps, gluteus medius (unseen)

Body drive: quadriceps, calves

Body twist: latissimus dorsi (unseen), abdominals

Key Static Muscles

Tibialis anterior

Primary Kinetic Chains

Lateral, hip turn, leg extension

Side angle
Strengthens quads; stretches legs, hips, and sides

Pigeon
Stretches hips, quads, and groin

COMMENTS

1) Hip flexibility generally diminishes with age. While front and back kicks are less affected by this change, others such as roundhouse and side kicks are markedly affected. To counter this trend, make sure to dedicate time to regular stretching.

BACK KICK

This popular kick is one of the three classic martial arts kicks—front (page 50), side (page 60), and back—that is widely taught and has an incredible number of variations. The back kick is the strongest of the three because it uses the powerful back and butt muscles; however, it doesn't have the accuracy of the other two kicks.

Speed (5 of 10)

The heavy musculature used for this kick makes it slower than others, but this relative lack of speed is often masked by the larger motion of a body turn.

Power (9 of 10)

The majority of power comes from the hip turn and the drive of both the kicking and supporting legs. Another important contributor is the tensing of the torso muscles, which adds body weight to the kick. The turning and jumping-turning back kicks are two common variations that are significantly more powerful than the standing kick.

Accuracy (6 of 10)

Accuracy is important, but this kick's high degree of power makes targeting a bit less critical. The main targets are near the opponent's center of mass, where even a partially blocked kick will still inflict damage. Other targets range from the head down to the thigh. As with the side kick, many people get the kicking leg and hip to work well, but focus less on coordinating the supporting leg and body weight. For example, if your supporting leg is planted too close to the target, you'll be knocked backward upon impact because your supporting leg isn't in position to drive into the target.

KEY EXERCISES

Leg swing backward
Strengthens glutes and hamstrings; stretches hips

T + opposite toe touch (page 129)
Develops balance; strengthens legs and core

Reverse half-moon
Strengthens legs and glutes; stretches legs, hips, spine, and chest; improves balance

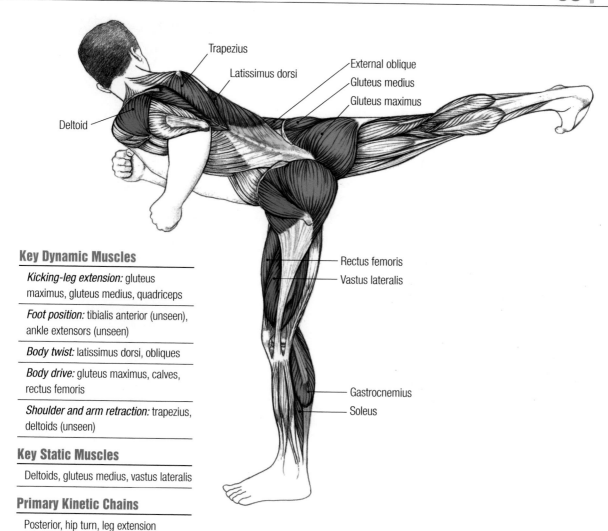

Trapezius

Latissimus dorsi

External oblique

Gluteus medius

Gluteus maximus

Deltoid

Rectus femoris

Vastus lateralis

Gastrocnemius

Soleus

Key Dynamic Muscles

Kicking-leg extension: gluteus maximus, gluteus medius, quadriceps

Foot position: tibialis anterior (unseen), ankle extensors (unseen)

Body twist: latissimus dorsi, obliques

Body drive: gluteus maximus, calves, rectus femoris

Shoulder and arm retraction: trapezius, deltoids (unseen)

Key Static Muscles

Deltoids, gluteus medius, vastus lateralis

Primary Kinetic Chains

Posterior, hip turn, leg extension

Wide-leg forward bend + shoulder stretch
Stretches hamstrings, adductors, and shoulders

Plow
Stretches shoulders and spine

COMMENTS

1) Shifting the line of the body changes the target of the kick. While sport-based arts often emphasize higher targets, lower targets are easier to teach and have more self-defense applications.

JUMPING ROUNDHOUSE KICK

The jumping roundhouse, jumping scissors roundhouse, and jumping-turning roundhouse kicks are common variations of the classic kick. These advanced kicks require a great deal of practice not only to master the kicks but also to learn when and how to use them against an opponent. Here we describe the back-leg jumping roundhouse kick.

Speed (7 of 10)

Speed is ultimately a function of the hip turn and the snapping out of the foot. However, many other factors, such as the countertorque generated by the shoulders and arms, assist in increasing speed.

Power (8 of 10)

This kick gains additional power from an upward jump followed by a twist, both of which are not utilized during a standard roundhouse kick.

Accuracy (6 of 10)

The coordination of five kinetic chains is an indicator of the complex interaction of the body in this kick, and this is in addition to the coordination of the initial jump. A clean, powerful set of jumping roundhouse kicks is a common goal of many advanced martial artists, and it takes an unusually long time to develop and maintain.

KEY EXERCISES

Jump with 180/360-degree turn (page 129) Improves lower-body explosive power and twist control

One-leg hop Improves lower-body explosive power

Band leg abduction Strengthens abductors

Key Dynamic Muscles

Jump (not pictured): quadriceps, calves

Kicking-leg extension: quadriceps

Kicking-leg lateral lift: gluteus medius (unseen), tensor fascia latae (unseen)

Body twist: pectorals, obliques

Key Static Muscles

Calves, rectus abdominis, platysma/sternocleidomastoid

Primary Kinetic Chains

Posterior, lateral, hip turn, shoulder turn, leg extension

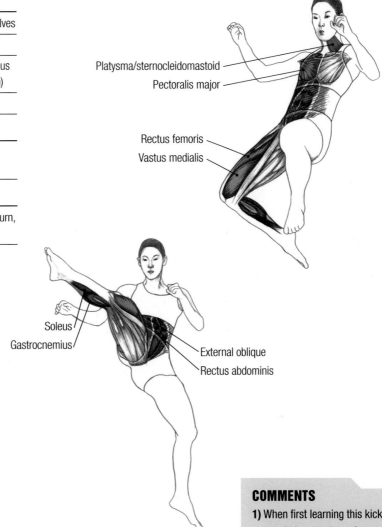

Platysma/sternocleidomastoid

Pectoralis major

Rectus femoris

Vastus medialis

Soleus

Gastrocnemius

External oblique

Rectus abdominis

Triangle
Strengthens quads; stretches legs, hips, shoulders, chest, and spine

Warrior 2
Strengthens legs, hips, and shoulders; stretches adductors

COMMENTS

1) When first learning this kick, most people learn to jump forward into it; others also learn it with a backward jump to create more space so that the kicking leg is not jammed by a charging opponent. In any case, be careful when landing—with the combination of jumping and twisting, it's not unusual to twist and damage the supporting leg.

2) As with all jumping techniques, be aware that being in the air makes you vulnerable to counterattacks.

SPINNING HEEL KICK

This visually pretty competition kick is unusually strong. Due to its somewhat long delivery time, it's usually used as a counterkick. The hard spin, usually away from an attack, makes this counterkick's point of attack unusual and thus hard to block. The vulnerability of the supporting leg and groin, however, makes this kick less common in self-defense situations and in competitions where leg and/or groin attacks are legal.

Speed (9 of 10)

Speed is of utmost importance, since the window of opportunity is very short; a mistimed spinning heel kick can leave you in a poor position to defend against a counterattack.

Power (7 of 10)

The spinning of the body generates the majority of power, with the arms and then the body twist contributing to the spin's power. The kicking leg is not quite straight (hyperextension of the kicking leg, especially when kicking boards, is a common injury for beginners), and initially there's substantial bend at the hip. Roughly 45 degrees before impact, the kicking leg accelerates through the target with a sharp but strong leg extension at the hip.

Accuracy (5 of 10)

This kick requires a lot of practice, especially in sustaining a balanced spin. The spin requires a strong lateral bend in the body because the kicking leg is held at the required height using a combination of the gluteus medius and the centripetal force of the spin.

KEY EXERCISES

Leg swing backward
Strengthens glutes and hamstrings; stretches hips

Reverse triangle
Strengthens legs; stretches legs, hips, spine, and chest

Reverse half-moon
Strengthens legs and glutes; stretches legs, hips, spine, and chest; improves balance

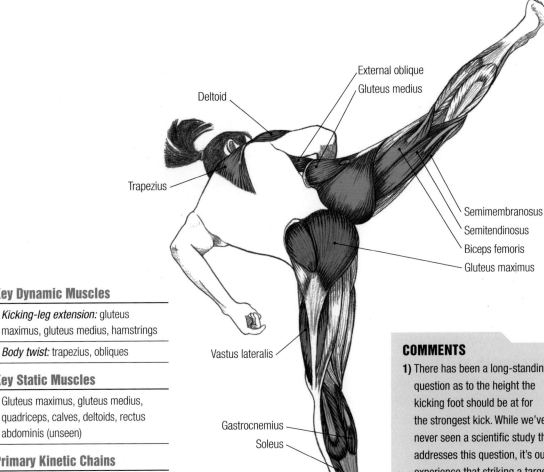

Deltoid

External oblique

Gluteus medius

Trapezius

Semimembranosus

Semitendinosus

Biceps femoris

Gluteus maximus

Vastus lateralis

Gastrocnemius

Soleus

Key Dynamic Muscles

Kicking-leg extension: gluteus maximus, gluteus medius, hamstrings

Body twist: trapezius, obliques

Key Static Muscles

Gluteus maximus, gluteus medius, quadriceps, calves, deltoids, rectus abdominis (unseen)

Primary Kinetic Chains

Posterior, lateral, shoulder turn, hip turn

Side crunch
Strengthens obliques

Pigeon
Stretches hips, quads, and groin

COMMENTS

1) There has been a long-standing question as to the height the kicking foot should be at for the strongest kick. While we've never seen a scientific study that addresses this question, it's our experience that striking a target at about mid-height (slightly above the hip) seems to generate the most power. If true, this may be due to the fact that, unlike the lower and higher kicks, the core body muscles are more optimally placed for power at the mid-level. This goes back to the old "peanut butter jar" theory of martial arts power: If you're handed a peanut butter jar and told the lid's stuck but needs opening, you'll usually place the jar in front of your navel (rather than, say, over your head) in order to exert maximum power on the lid.

LOW SPINNING HEEL KICK

This kick is often used as a counterkick against an opponent who has moved in with a fast and/or strong attack, usually a kick. Dropping down to the ground and spinning make the point of attack quite unusual. If there's weight on the opponent's leg that's being attacked, it's difficult—if not dangerous—to withstand the kick.

Speed (9 of 10)
Speed is of utmost importance since the window of opportunity is very short; a mistimed low spinning heel kick can leave you in a poor position to defend against a counterattack.

Power (7 of 10)
The spinning of the body generates the majority of power, with the body twist contributing to the spin's power. The kicking leg is not quite straight (hyperextension of the kicking leg, especially when kicking boards, is a common injury for beginners), and there's substantial bend at the hip. Roughly 45 degrees before impact, the kicking leg accelerates through the opponent's legs with a sharp but strong leg extension.

Accuracy (5 of 10)
This kick requires lot of practice, especially in sustaining a balanced spin. The spin requires a strong lateral (side) bend in the body because the kicking leg is held off the ground mostly by the gluteus medius.

KEY EXERCISES

Band leg abduction
Strengthens abductors

Toe walk (page 129)
Strengthens calves

One-legged bridge + hip dip (page 129) Strengthens pelvic thrust; stretches chest and shoulders

Key Dynamic Muscles

Kicking-hip extension: gluteus maximus

Body twist: obliques

Key Static Muscles

Gluteus medius, hamstrings, calves (unseen), quadriceps (supporting leg, unseen)

Primary Kinetic Chains

Posterior, lateral, hip turn, shoulder turn

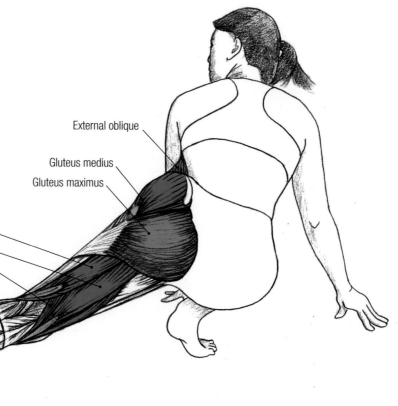

External oblique

Gluteus medius

Gluteus maximus

Biceps femoris

Semitendinosus

Semimembranosus

Side angle
Strengthens quads; stretches legs, hips, and sides

Pigeon
Stretches hips, quads, and groin

COMMENTS

1) Spinning on the ball of the supporting foot extends the kick's range, but balance is harder to maintain. Some people prefer to spin on the knee of the supporting leg because it's easier to balance, but this reduces the kicking range to about the length of the thigh. This variation is also quite rough on the knee because a fast drop onto the knee can cause serious injury.

THROWS

The throws featured in this section are fairly basic, yet this section was the most difficult to write due to the complexity of the average throw. While this book's premise is to highlight and describe the key muscles used in a given technique, a throw has a number of phases, each one of which uses a different set of key muscles. Due to space constraints, we will concentrate on the casting component of the throwing process, as opposed to the balance break or loading.

We chose a wide variety of throws to illustrate different features of basic throwing techniques. Some throws, such as the minor outer reap and the forward body drop, demand a high degree of accuracy. Others, such as the sweeping hip throw and the front fireman's throw, require more strength. The shoulder throw and forward body drop involve turning your back to your opponent, while throws such as the rice bale and the minor outer reap are done face-to-face. Finally, some, like the front fireman's throw, are done to the side.

At its simplest, an "average" throw might be broken down into three parts: the balance break of the opponent, the loading, and the casting. For each of the nine throws shown in this section (although the snapover is a finish for several throws rather than an actual throw itself), we'll highlight the casting portion of the throw; occasionally, we'll acknowledge key muscles for preliminary movements, such as arm pulls involved in balance breaking. However, it's not unusual for other aspects of the technique to be difficult to master or require extensive exercises to work up to. For example, the pickups for the front fireman's throw and rice bale throw require strength, balance, and timing; the finish of a forward body drop also demands strength, balance, and timing as the thrown opponent is snapped over. These aspects are at most only mentioned in this section.

Throws

- Stomach Throw

- Major Outer Reap

- Minor Outer Reap

- Forward Body Drop

- Shoulder Throw

- Snapover

- Sweeping Hip Throw

- Rice Bale Throw

- Front Fireman's Throw

STOMACH THROW

While this throw can be executed as an offensive technique, more often than not it's used as a defense against a lunging or charging opponent.

Speed (4 of 10)

When a stomach throw is executed as an offensive technique, at least for the throw entry, high speed is usually necessary to get under the opponent's weight. When executed as a defensive technique, the move is often slower since it needs to match the attacker's speed and power. This blend of speed and power allows you to use the attacker's strength against him.

Power (7 of 10)

The majority of the power is generated by two aspects of the throw:

Forward pull: Unless the opponent is in a headlong charge, you'll have to pull your opponent forward to break his balance, allowing you to slide under his mass.

Leg extension: With your opponent's balance broken, your extending leg lifts his body off the ground. You can accentuate the leg extension by thrusting your hips.

Accuracy (8 of 10)

Done correctly, this throw is circle-shaped, with the bodies of the thrower and opponent forming the outer ring and the extending leg serving as a spoke; this is why it's sometimes referred to as a circle throw. The body twist, a combination of lying under your opponent at an angle and pulling him forward and over your opposite shoulder as you extend your leg, ensures that your opponent doesn't come down on you. These components change in importance depending on the throwing situation.

KEY EXERCISES

One-arm dumbbell row
Strengthens trapezius

Barbell/dumbbell pullover
Strengthens pecs, triceps, and lats

Bicycle crunch
Improves core flexibility and strength

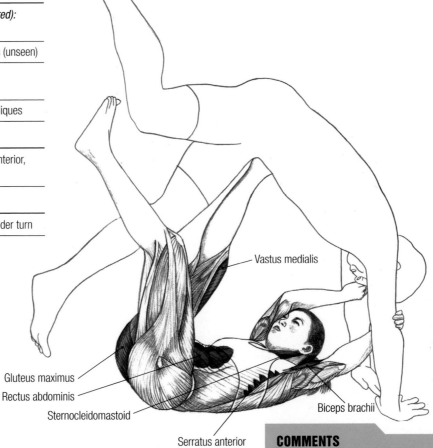

Key Dynamic Muscles

Body pull forward (not pictured): pectorals, biceps, deltoids

Forward pull: biceps, deltoids (unseen)

Leg extension: quadriceps, gluteus maximus

Body twist (not pictured): obliques

Key Static Muscles

Rectus abdominis, serratus anterior, sternocleidomastoid

Primary Kinetic Chains

Posterior, leg extension, shoulder turn

Vastus medialis

Gluteus maximus

Rectus abdominis

Sternocleidomastoid

Serratus anterior

Biceps brachii

One-legged bridge + hip dip (page 129)
Strengthens pelvic thrust; stretches chest and shoulders

Roll around (page 129)
Warms up back and hips

COMMENTS

1) This throw is sometimes referred to as the "Captain Kirk throw" due to the number of times the *Star Trek* character executed the technique during the filming of the show.

2) The most common mistake for beginners is that they pull their opponent down on top of themselves instead of pulling the opponent past them. Some instructors refer to this as "getting a flat tire" since this changes the correct circular motion into a flat, linear motion.

MAJOR OUTER REAP

Perhaps the simplest and safest throw to teach a beginner, the major outer reap is often one of the first throws that is taught. This throw is usually most effective as a secondary throw or a counter to a throw.

Speed (6 of 10)

When used as a countertechnique, the speed of the throw is often dictated more by the attacker's speed and power than anything else. For example, when an opponent sharply pulls on you, you need to have a fast, strong entry into the throw that blends with the opponent's pulling action.

Power (8 of 10)

Two key factors in power generation are:

Body twist: In this throw, you strongly close on your opponent while pulling him in and to the side until the corners of your shoulders meet. Upon impact, one arm pulls as the other arm pushes, thus twisting your opponent and breaking his balance.

Leg reap: During the shoulder twist, your inside leg, bent at a slight angle, reaps through either one or both of your opponent's legs. This reap is not only powered by the leg swing, it's also driven by the forward pitch of your body.

Accuracy (5 of 10)

Closing the distance between your body and your opponent's is critical. Trying to reap someone who is as little as a few inches away can greatly diminish this technique's effectiveness. It's also very important to reap or chop all the way through your opponent's legs. Beginners often don't reap far enough, and this gives the opponent a chance to regain lost balance instead of being thrown.

KEY EXERCISES

Leg swing backward
Strengthens glutes and hamstrings; stretches hips

Lunge + twist
Enhances hip flexibility while developing core power

Woodchopper
(page 129)
Strengthens obliques and shoulders

Key Dynamic Muscles

Body drive: quadriceps (unseen), calves

Body pull-in and twist: pectorals (unseen), biceps (unseen), deltoids, latissimus dorsi, obliques, rectus abdominis (unseen)

Leg reap: gluteus maximus, hamstrings

Key Static Muscles

Calves

Primary Kinetic Chains

Posterior, shoulder turn

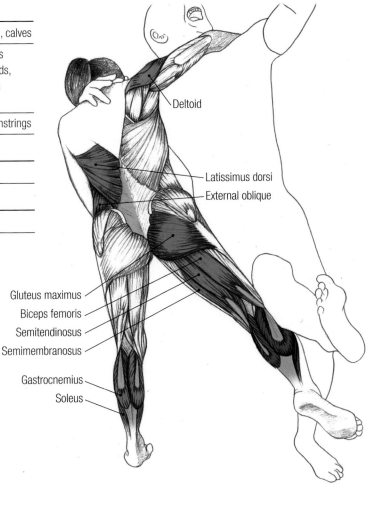

Deltoid

Latissimus dorsi

External oblique

Gluteus maximus

Biceps femoris

Semitendinosus

Semimembranosus

Gastrocnemius

Soleus

Inchworm (page 128)
Strengthens arms, shoulders, pecs, and core; stretches hamstrings

Reverse half-moon
Strengthens legs and glutes; stretches legs, hips, spine, and chest; improves balance

COMMENTS

1) It's not uncommon for this throw to be done so hard and fast that the thrower ends up rolling forward and past their falling opponent.

2) Driving with the supporting leg is critical to breaking your opponent's balance.

MINOR OUTER REAP

This close-contact throw requires timing and full-body commitment to successfully accomplish. It doesn't rely heavily on power, which makes it all the more important to learn the careful timing when trying to master this exacting move.

Speed (7 of 10)

Speed is required to time the move with the opponent's movement. The out-to-in reap is not very powerful and thus its success relies on speed and timing.

Power (3 of 10)

Power is generated primarily by the arm movements. The arm of the sweeping leg pulls backward while the other arm pushes forward, all in an effort to twist the opponent off balance. The legs also make opposite motions: The front leg reaps from outside to inside and the back leg straightens as it drives hard into the opponent.

Accuracy (8 of 10)

At the risk of being too general, you could say that the less powerful the throw, the more important the timing becomes. The minor outer reap is the poster child for this statement. The twisting of both arms and legs, as well as the timing of these movements with your opponent's, make this throw one of the more difficult to master.

KEY EXERCISES

Squat with partner (page 129)
Strengthens quads and glutes

Bicycle crunch
Improves core flexibility and strength

Clapping push-up (page 128)
Improves explosive upper-body power

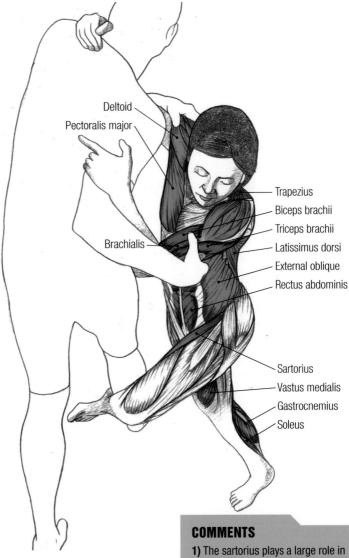

Deltoid

Pectoralis major

Brachialis

Trapezius

Biceps brachii

Triceps brachii

Latissimus dorsi

External oblique

Rectus abdominis

Sartorius

Vastus medialis

Gastrocnemius

Soleus

Key Dynamic Muscles

Arm pull (left arm, as illustrated): deltoids, trapezius

Arm push (right arm, as illustrated): pectorals, triceps (unseen), deltoids

Shoulder twist and lateral drive: obliques, latissimus dorsi

Leg sweep: sartorius, hamstrings (unseen)

Body drive: quadriceps, calves

Key Static Muscles

Rectus abdominis, biceps, brachialis, triceps

Primary Kinetic Chains

Posterior, lateral, leg extension, shoulder turn

Tree
Improves balance; stretches and strengthens legs and hips

Kneeling lunge
Stretches hip flexors and quads

COMMENTS

1) The sartorius plays a large role in moving the reaping leg, but it's a relatively weak muscle and cannot execute the throw without help—it requires body movement and the arms pushing and pulling to break the opponent's balance and to get their weight at least partially off their reaped leg.

2) To coordinate the throw, think of your hands and your reaping foot powering around the outside edge of a large circle.

FORWARD BODY DROP

Often referred to as a hand throw (or *te waza* in Japanese), the forward body drop doesn't rely on much power, so both speed and accuracy are essential to successfully execute this technique. As with other less-powerful moves, this is usually a defensive or responsive throw is executed by redirecting the opponent's power.

Speed (9 of 10)

Speed is essential to blend the motion of your body with that of your opponent's. This throw involves a large amount of motion: Your hips and body must be turned and your arms must be extended to guide your opponent forward. Once he loses his balance, your hands must pull in sharply to snap him over (this snapover is described on page 82).

Power (4 of 10)

The majority of this throw's power comes from the opponent's forward movement, which is redirected and amplified to break his balance and produce the throw. While the power you add is not great, it's necessary, and its timing is critical.

Accuracy (8 of 10)

Blending your body movement with your opponent's requires an unusually high degree of commitment, since you turn your back on him and leave yourself open to a counterattack if the technique fails.

KEY EXERCISES

Clapping push-up (page 128)
Improves explosive upper-body power

Standing band pull (page 129)
Strengthens traps, triceps, delts, serratus anterior, pecs, and abs

Crunch (feet up)
Strengthens core muscles

Key Dynamic Muscles

Initial arm pull (not pictured): pectorals, deltoids, biceps, brachioradialis

Arm extension (left arm, as illustrated): posterior deltoid, triceps

Body twist: obliques (unseen), pectorals

Body drive: quadriceps, calves

Key Static Muscles

Rectus abdominis, gluteus maximus (supporting leg, unseen), anterior deltoid, biceps

Primary Kinetic Chains

Posterior, hip turn, shoulder turn, arm extension

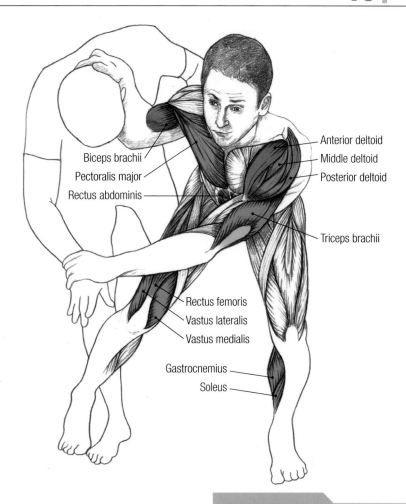

Biceps brachii
Pectoralis major
Rectus abdominis

Anterior deltoid
Middle deltoid
Posterior deltoid

Triceps brachii

Rectus femoris
Vastus lateralis
Vastus medialis

Gastrocnemius
Soleus

High-elbow shoulder stretch
Stretches shoulders and triceps

Rear palm press
Stretches wrists and forearms

COMMENTS

1) Once the opponent is falling forward, the sharpness of the fall can be increased by pulling the extended arms back in.

2) The placement of the main throwing hand on or around the opponent's head is quite varied. Some place their hand behind the neck, which makes the forward pull easier. However, others push their hand up under the chin to help break their opponent's balance; they then slip their hand behind the neck for the forward pull.

SHOULDER THROW

The shoulder throw is considered one of the more basic throws after perhaps the hip throw and major outer reap (page 74). While the shoulder throw and its many variations are of primary importance in competition, in self-defense classes it's taught more as a defense against rear attacks than as an offensive move due to the issues involved in turning your back on your opponent.

Speed (5 of 10)

The shoulder throw can be quite fast when it's used as an offensive technique. However, as a defensive move, its speed is often dictated by the attacker's speed and power, since the throw usually blends in with the speed of the opponent's attack.

Power (7 of 10)

The power inherent in the shoulder throw comes in two phases: popping the opponent off the ground using the posterior and leg extension kinetic chains, and twisting the shoulders, and to a lesser extent the hips. Perhaps the single most common mistake that lessens the technique's power is letting the throwing shoulder get too far in front of the opponent's body, which makes the shoulder twist less forceful, if not impossible to perform. Tight body contact between you and your opponent is key to this move's efficiency.

Accuracy (6 of 10)

A wide variety of shoulder throws are taught, including one-arm, two-arm, dropping, and leg-assist variations. Some emphasize speed and others power, and it's a matter of training and experience to choose which one to use depending on the relative speed, power, and position of the opponent's attack. With all these variations, one of the most important points to remember is to make sure that your body is low and properly aligned in front of your opponent before executing the initial body pop.

KEY EXERCISES

Squat with partner (page 129) Strengthens quads and glutes

Woodchopper (page 129) Strengthens obliques and shoulders

Standing band pull (page 129) Strengthens traps, triceps, delts, serratus anterior, pecs, and abs

Key Dynamic Muscles

Initial body pull-in (not pictured):
pectorals, biceps, deltoids

Arm extension (left arm as illustrated):
deltoids, triceps

Shoulder twist: pectorals, obliques
(unseen), rectus abdominis

Leg extension: quadriceps, calves

Key Static Muscles

Gluteus maximus (unseen), biceps,
brachialis

Primary Kinetic Chains

Posterior, leg extension, shoulder turn

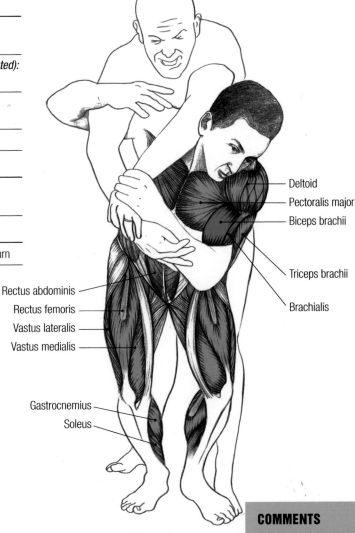

Deltoid
Pectoralis major
Biceps brachii
Triceps brachii
Brachialis

Rectus abdominis
Rectus femoris
Vastus lateralis
Vastus medialis

Gastrocnemius
Soleus

Lunge + twist
Enhances hip flexibility while
developing core power

High-elbow shoulder stretch
Stretches shoulders and triceps

COMMENTS

1) Take special care to not let your
throwing arm reach behind the
plane of your chest since it weakens
your ability to pull and a resisting
opponent can yank back on the arm
and dislocate your shoulder.

2) The double-arm shoulder throw
(shown) tends to give your
opponent an arm bar as they're
being thrown. Be careful when
practicing this variation as elbow
damage can easily occur.

SNAPOVER

This transitional technique is commonly used in conjunction with an initial technique such as a hip or wrist throw, and is then followed up perhaps with a finish-off (e.g., wrist lock, arm lock, or stomp kick). The most important aspect of this technique is that as you throw your opponent and he is falling freely, you must sharply pull in your arms, which suddenly snaps the faller over. This action accentuates the fall's impact and puts you into a good position to perform a finishing technique.

Speed (6 of 10)

The technique's inward spiral generates its speed. Most throws (e.g., hip, shoulder, and wrist) require a large initial arc of your body as your opponent's balance is broken and the throw is initiated. Once your opponent is in the air, he becomes very easy to manipulate, but only for a very short window of time; this means you need to exactly time the snapover with your body's movement.

Power (8 of 10)

The majority of power comes from two sources: The front leg extension pushes your body backward, which starts the snapover of your falling opponent; and the inward arm pull, with major assistance from the back muscles, finishes the move and adds crispness to the throw.

Accuracy (6 of 10)

The timing of this snapover action is difficult to master. Starting the action too early or too late will negate the effectiveness of the technique and could even put you into a precarious position that's vulnerable to counterattack.

KEY EXERCISES

Burpie
Improves full-body explosive power

Body drag—pull (page 128)
Strengthens traps, lats, and quads

T + opposite toe touch (page 129)
Develops balance; strengthens legs and core

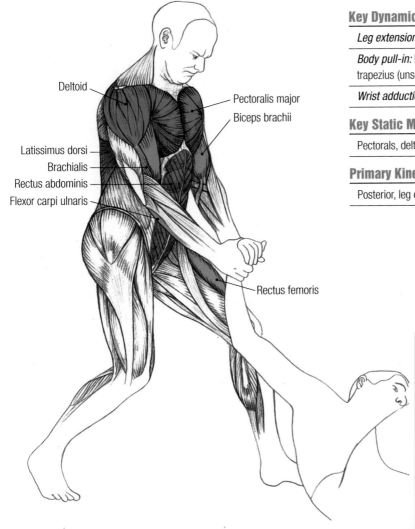

Deltoid

Pectoralis major

Biceps brachii

Latissimus dorsi

Brachialis

Rectus abdominis

Flexor carpi ulnaris

Rectus femoris

Key Dynamic Muscles

Leg extension: quadriceps

Body pull-in: biceps, brachialis, trapezius (unseen), latissimus dorsi

Wrist adduction: wrist adductors

Key Static Muscles

Pectorals, deltoids, rectus abdominis

Primary Kinetic Chains

Posterior, leg extension

COMMENTS

1) Note that this is a transitional technique. It's assumed that the thrower will be moving into any one of a number of finish-off positions, including pulling up on the wrist and stomp-kicking the ribs; twisting the arm to produce a wrist, elbow, or shoulder lock; or twisting the arm to force the opponent onto their face and into any number of submission or hold positions.

2) To prevent injuries and reduce the chance of counterattacks, always push your hips forward and lean back so your entire body, from legs up through arms, participates in the final pull. Leaning forward and rounding your back can lead to back injuries and makes you vulnerable to being pulled down onto your opponent.

Push-up + one-arm row
Strengthens core, lats, traps, and delts

Reverse plank
Stretches arms, shoulders, and front of body

SWEEPING HIP THROW

This is arguably the most powerful of the numerous hip-throw variations because the sweeping leg virtually spins the opponent into the ground. Perhaps the biggest difficulty in executing the throw comes from the fact that as you sweep your leg, you're standing on only one leg, which puts a premium on body placement.

Speed (6 of 10)

The speed of this technique is no faster than a regular hip throw, but due to the spin of the opponent, the speed of the fall is greater.

Power (9 of 10)

The power inherent in the sweeping hip throw comes in three phases: popping your opponent off the ground, sweeping the leg, and twisting the shoulders.

Accuracy (6 of 10)

Body alignment is key due to the momentary balance and drive that must be accomplished with the one supporting leg. Note that the word "balance" in this case does not refer to a lack of motion, as in a static position; it refers to the control under which this very dynamic throw is being held throughout its execution.

KEY EXERCISES

Leg swing backward
Strengthens glutes and hamstrings; stretches hips

Woodchopper (page 129) Strengthens obliques and shoulders

Standing band pull (page 129) Strengthens traps, triceps, delts, serratus anterior, pecs, and abs

Key Dynamic Muscles

Supporting-leg extension: quadriceps, calves

Body pull-in & twist: pectorals, deltoids, obliques, rectus abdominis

Arm pull (left arm, as illustrated): trapezius, deltoids (unseen), triceps

Leg sweep: gluteus maximus (unseen), hamstrings (unseen)

Key Static Muscles

Gluteus maximus (supporting leg, unseen), quadriceps, calves, trapezius, posterior deltoid, biceps (unseen)

Primary Kinetic Chains

Posterior, leg extension, shoulder turn

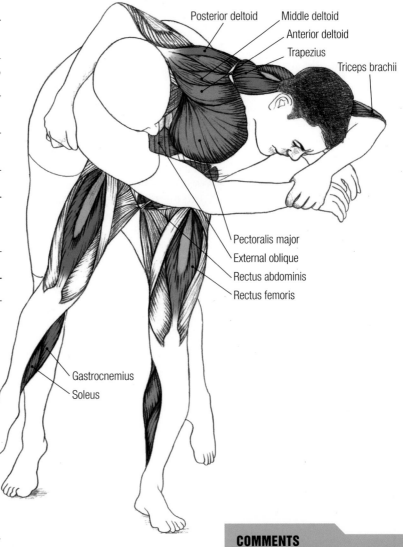

Posterior deltoid
Middle deltoid
Anterior deltoid
Trapezius
Triceps brachii
Pectoralis major
External oblique
Rectus abdominis
Rectus femoris
Gastrocnemius
Soleus

Side crunch
Strengthens obliques

Reverse half-moon
Strengthens legs and glutes; stretches legs, hips, spine, and chest; improves balance

COMMENTS

1) It's not uncommon for the thrower to drive himself off his own feet during the execution of this throw. Learning to do this safely is obviously an important aspect of learning this throw.

2) Take care not to let your inside arm (right, as illustrated) get behind the plane of your chest because shoulder injuries, including dislocation, can occur with a resisting opponent.

RICE BALE THROW

The rice bale throw's colorful but apt name comes from the way you're supposed to safely pick up a heavy bale of rice. While wrestlers, MMA practitioners, and judoka know this throw, it's not used often in self-defense situations, as it requires grabbing an opponent between the legs.

Speed (4 of 10)

Speed is important but really only becomes critical at the moment of popping the opponent off the ground and completing the throw. This technique often presents itself as a momentary opportunity during a body clash—for example, you manage to slip an opponent's punch and get inside, and then find that as your bodies slam into each other, your opponent has straightened up for a second. Speed becomes essential at this moment to execute the throw.

Power (8 of 10)

Surprisingly little power is required to execute the lifting portion of this move. The more important power stroke is pulling the opponent in just before the lift. As you lower your center of mass below your opponent's and pull him into your hips, there's a moment when straightening your legs is all that's required to get your opponent into the air. Once the opponent is airborne, and with both your centers of mass (usually just below the navel) aligned, it's simply a matter of rotating him 90 degrees for the fall. For demos, it's not uncommon to spin an opponent 270 degrees for a fall, or even 360 degrees and back to their feet.

Accuracy (6 of 10)

The most important issue with this throw is in choosing when to attempt it. The combatants must be quite close together, with their centers of mass (i.e., their hips) virtually touching. The thrower's hips should be below the opponent's as this greatly reduces the amount of power required.

KEY EXERCISES

Burpie
Improves full-body explosive power

Deadlift (page 128)
Strengthens glutes, quads, and traps

Upright row
Strengthens traps, delts, brachialis, and brachioradialis

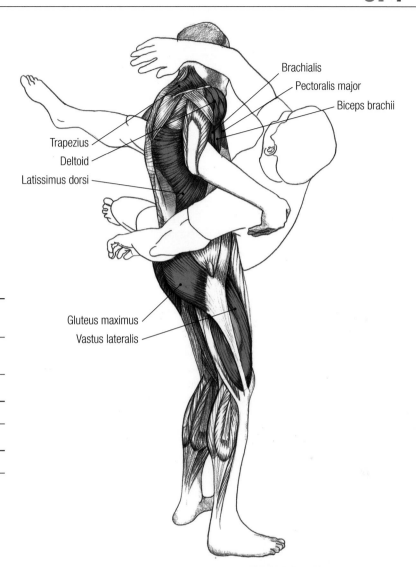

Brachialis
Pectoralis major
Biceps brachii
Trapezius
Deltoid
Latissimus dorsi
Gluteus maximus
Vastus lateralis

Key Dynamic Muscles

Body pull-in: pectorals, biceps, brachialis, deltoids, latissimus dorsi

Body thrust and lift: gluteus maximus, quadriceps, trapezius

Key Static Muscles

Rectus abdominis (unseen)

Primary Kinetic Chains

Posterior, leg extension

Wide-leg forward bend + shoulder stretch
Stretches hamstrings and shoulders

Child's pose
Stretches hips, quads, back, and shoulders

COMMENTS

1) Severe back injuries can occur if you try to pick up an opponent when your back is not straight and when you use your back (rather than your legs).

2) Given the importance of lining up the centers of mass, note that due to differences in build, the average man's center of mass is higher than that of the average woman.

FRONT FIREMAN'S THROW

Usually used as a defensive as opposed to an offensive technique, the fireman's throw has two basic entries, the front and back. There's also a huge number of variations that include throwing from standing or from the knees, and throwing the opponent laterally or over the head and to the front of the body. Some versions have the opponent go over the shoulders, while others have the opponent rotate over the hips. The illustrated throw shows the classic lateral throw that goes across the shoulders.

Speed (4 of 10)

The speed of this throw is somewhat dictated by the speed and power of the opponent's attack. The faster the attack, the faster the throw must be in order to blend with the opponent's momentum.

Power (7 of 10)

Lifting your opponent off the ground in this move requires the most power. However, you can argue that by blending with the opponent's attack, the power requirement is minimized because the opponent's power does most of the work. Unfortunately, this kind of perfect blending is more the exception than the rule, and you must be prepared to add power to the throw to offset the usual imperfections. It's often more practical for smaller throwers to throw an opponent over their hips rather than their shoulders.

Accuracy (8 of 10)

Blending with your opponent's forward momentum is critical. While it's usually true that you must choose an appropriate throw that blends with your opponent's movement, it's especially important with the fireman's throw—a poorly chosen moment to attempt this throw exposes your head and neck to counterattack.

KEY EXERCISES

Squat with partner (page 129)
Strengthens quads and glutes

Warrior 2 band pull (page 129)
Strengthens legs, hips, shoulders, and triceps; stretches chest

Side crunch
Strengthens obliques

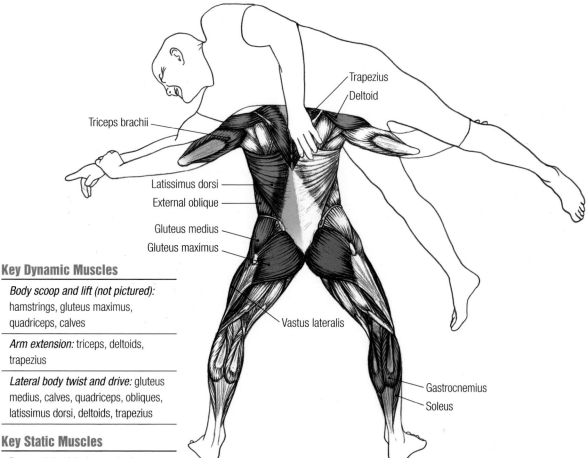

Trapezius

Deltoid

Triceps brachii

Latissimus dorsi

External oblique

Gluteus medius

Gluteus maximus

Vastus lateralis

Gastrocnemius

Soleus

Key Dynamic Muscles

Body scoop and lift (not pictured): hamstrings, gluteus maximus, quadriceps, calves

Arm extension: triceps, deltoids, trapezius

Lateral body twist and drive: gluteus medius, calves, quadriceps, obliques, latissimus dorsi, deltoids, trapezius

Key Static Muscles

Rectus abdominis (unseen), gluteus maximus, quadriceps

Primary Kinetic Chains

Lateral, leg extension, arm extension

High-elbow shoulder stretch
Stretches shoulders and triceps

Rear palm press
Stretches wrists and forearms

COMMENTS

1) The deltoids only raise the arms up to a horizontal position; after that, the trapezius takes over as it rotates the shoulder blades to continue raising the arms higher. These interwoven mechanisms of the deltoids and the trapezius are key in the loading and throwing portions of fireman's throws.

2) To prevent injury, do not round your back during the initial lifting phase of the throw. Throw larger partners over the small of the back rather than over the shoulders.

GROUNDWORK

Groundwork involves taking down an opponent and manipulating them on the ground. While there is certainly a wide variety of offensive techniques (like strikes, pressure points, chokes, and joint locks) that can be done while you're on the ground, this section covers just the absolute basics of groundwork.

Of the six techniques shown, two involve taking down a standing opponent while the other four focus on holding an opponent on the ground. These techniques do not use a great deal of strength; rather, they involve shifting weight and applying pressure at key points.

While proper technique will help tremendously in holding larger and stronger opponents, note that when the weight and strength differential between two opponents increases, the number of effective and practical techniques becomes quite limited. In some competitions, a 10 percent difference in weight can be used to separate weight classes; in reality, the use or misuse of weight can be an asset or a liability.

Groundwork

- **Guard**

- **Scarf Hold**

- **Side Mount**

- **Bridge & Shrimp**

- **Single-Leg Takedown**

- **Double-Leg Takedown**

GUARD

This defensive position uses the length of the torso and the strong core muscles to keep a larger opponent at bay. By holding your opponent with your legs, you free your arms for both offense and defense.

Speed (2 of 10)

This relatively static position requires little movement and thus little speed. However, as your opponent tries to escape or attack, you'll need to react quickly with a technique that is separate but works in conjunction with the guard position. For example, if your opponent reaches forward with one arm, you can quickly abandon the guard position in favor of an arm bar.

Power (7 of 10)

Power is generated primarily from the legs and torso and is used to keep the opponent at arm's length. By arching and twisting your body, you can execute different attacks and defenses. There's a huge variety of moves and countermoves from this position.

Accuracy (5 of 10)

This position's static nature can change rapidly as the opponent moves. For example, your opponent may lean forward to strike or twist to the side to try to escape; the precision of your countertechniques will determine their success.

KEY EXERCISES

Band leg adduction
Strengthens adductors

Supine leg push-down (page 129)
Improves core strength

V-up
Improves core strength

Key Dynamic Muscles

Leg wrap and squeeze: adductors, gracilis, pectineus, sartorius

Body extension: quadratus lomborum (unseen)

Key Static Muscles

Gluteus medius (unseen), rectus abdominis

Primary Kinetic Chains

Posterior

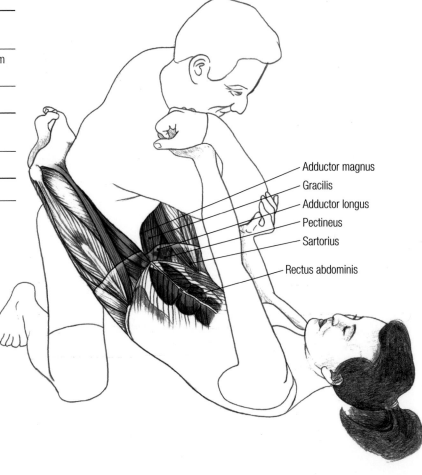

Adductor magnus
Gracilis
Adductor longus
Pectineus
Sartorius
Rectus abdominis

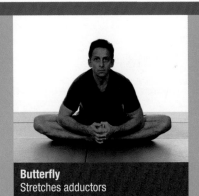

Butterfly
Stretches adductors

Bow
Extends the body

COMMENTS

1) In recent years, the guard has been a popular position to teach because a variety of techniques are associated with it. People in self-defense classes sometimes argue that this position should be avoided because being in guard opens you up to groin attacks.

SCARF HOLD

This popular side hold is a good position from which other techniques, such as arm bars and chokes, can be executed. This hold is used extensively in competition, but is not often taught in self-defense classes due to its limited flexibility and its limitations when an opponent is very large.

Speed (4 of 10)

Speed is not critical except when responding to an opponent's attempts to escape; recognizing those attempts early will allow time for countermeasures.

Power (8 of 10)

Power comes from the core muscles, body weight, and leg drive. The side of your chest must rest on the corner of your opponent's chest; tensing your chest muscles will concentrate the force on as small an area as possible. Other key factors include:

Head lock: Keeping a tight grip on your opponent's neck and shoulder is important for impeding his movements.

Arm pull: Pulling sharply on your opponent's arm generates a lot of tension, which further helps to impede movement.

Leg walk: Your legs must be kept out to the side, out of your opponent's reach. As your opponent struggles, "walking" your legs around helps to maintain your weight on your opponent.

Accuracy (6 of 10)

One of the keys to an effective scarf hold is to make sure that the line between your hips and solar plexus is at a right angle to the same line on the person you're holding. This is where the leg walk comes into play.

KEY EXERCISES

One-arm dumbbell row
Strengthens traps

Biceps curl
Strengthens biceps

Low side plank
Improves core strength and stability

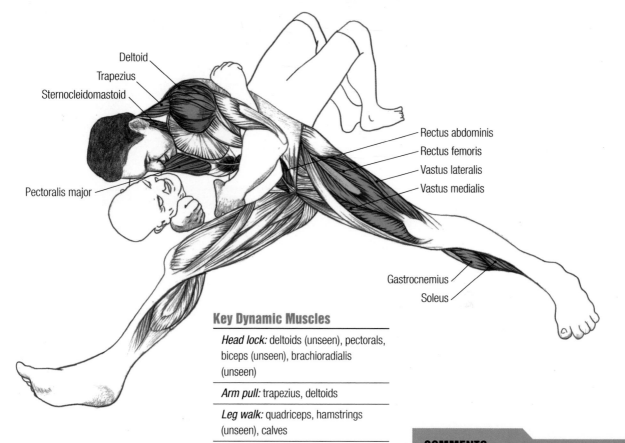

Deltoid
Trapezius
Sternocleidomastoid
Pectoralis major

Rectus abdominis
Rectus femoris
Vastus lateralis
Vastus medialis

Gastrocnemius
Soleus

Key Dynamic Muscles

Head lock: deltoids (unseen), pectorals, biceps (unseen), brachioradialis (unseen)

Arm pull: trapezius, deltoids

Leg walk: quadriceps, hamstrings (unseen), calves

Key Static Muscles

Abdominals, sternocleidomastoid

Primary Kinetic Chains

None

COMMENTS

1) While the scarf hold is relatively stable, it's not uncommon for the position to be compromised and thus force you to abandon it in favor of another hold, such as a side or top mount.

2) The arm around the neck (right arm as illustrated) can be used to attack the back of the opponent's neck by making a sawing action with the sharp edge of the forearm (the radius). This is an important aspect in keeping an opponent uncomfortable. A strong brachioradialis is essential for this and can be achieved by doing weight exercises such as hammer curls.

Locust
Stretches and strengthens the back body

Supine twist
Improves spine flexibility

SIDE MOUNT

The side mount is a reasonably strong pin, or holding position. It's also flexible, which allows you to easily and rapidly transition from one position to another to compensate for your opponent's weight shifts and body twists as he attempts to escape.

Speed (2 of 10)

This position is relatively static since there's little motion required except in response to your opponent's movements. While those responses need to be quick to be effective, the side mount itself requires mostly muscle tension and balance shifts.

Power (6 of 10)

Shifting your weight in response to your opponent's movements generates most of the power required for this technique. Tensing your muscles as you push your body weight sharply into key locations (such as the chest or hips) on your opponent's body keeps your opponent on the ground and under control. Arching your back, pulling in with your arms, and pushing inward with your feet also keep your weight and tension on your opponent.

Accuracy (8 of 10)

Keeping a balanced distribution of weight and moving your body weight in response to your opponent's body shifts are the key components for maintaining accuracy.

KEY EXERCISES

Mountain climber
Improves lower-body power

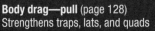

Body drag—pull (page 128)
Strengthens traps, lats, and quads

High shoot (page 128)
Strengthens arms and core; improves agility

Key Dynamic Muscles

Arm pull-in: trapezius, latissimus dorsi, biceps, brachialis

Hip extension: gluteus maximus

Leg drive: quadriceps, calves

Key Static Muscles

Deltoids

Primary Kinetic Chains

Posterior, lateral

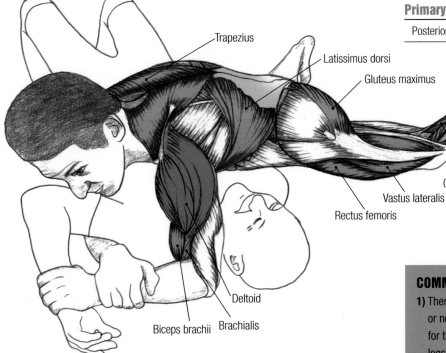

Trapezius

Latissimus dorsi

Gluteus maximus

Soleus

Gastrocnemius

Vastus lateralis

Rectus femoris

Deltoid

Brachialis

Biceps brachii

COMMENTS

1) There is some debate on whether or not the legs should be pulled in for this hold. Pulling one or both legs in can be more powerful in terms of immobilizing an opponent, but some argue that in this position the groin is closer to the opponent and thus more vulnerable to attack.

2) While over 90 percent of shoulder dislocations are anterior and inferior (forward and down), the figure-four lock shown above strongly twists the shoulder forward and up. Without due care, this can cause a dislocation in that direction.

Child's pose
Stretches hips, quads, back, and shoulders

Cobra
Stretches chest, shoulders, and abs

BRIDGE & SHRIMP

This combination move is an escape for when you're lying on your back with an opponent sitting on your hips or stomach. By popping up your hips into a bridge, you create enough space to thrust and twist (or "shrimp") your opponent off.

Speed (5 of 10)

Speed is important but overshadowed by the timing of this technique. While it's important to bridge quickly to pop your opponent's weight up and off, it's the speed of the subsequent shrimping action that allows you to escape.

Power (8 of 10)

The main burst of power is during the bridge, an upward pelvic thrust that disrupts your opponent's weight. The shrimping action starts when you thrust out your arms to continue your opponent's momentum up and over your head. This is immediately followed by a sharp hip twist.

Accuracy (6 of 10)

The coordination of the hip thrust, arm extension, and body twist must be fairly accurate or else the escape will fail and you may find yourself in an even worse position than the one in which you started.

KEY EXERCISES

Low shoot (page 129)
Strengthens arms and core; improves agility

Barbell/dumbbell pullover
Strengthens pecs, triceps, and lats

One-legged bridge + hip dip (page 129)
Strengthens pelvic thrust; stretches chest and shoulders

Key Dynamic Muscles

Pelvic thrust: quadriceps, gluteus maximus

Arm deflection and twist: deltoids, triceps, anconeus, serratus anterior

Shrimp & twist: obliques (unseen), calves, latissimus dorsi (unseen), pectorals

Key Static Muscles

Rectus abdominis, pectorals

Primary Kinetic Chains

Posterior, hip twist, shoulder twist

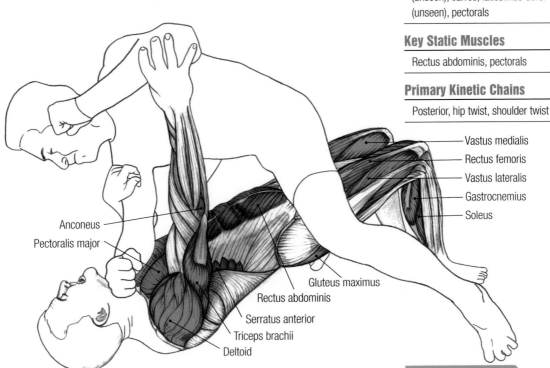

Vastus medialis
Rectus femoris
Vastus lateralis
Gastrocnemius
Soleus

Anconeus
Pectoralis major

Gluteus maximus
Rectus abdominis
Serratus anterior
Triceps brachii
Deltoid

COMMENTS

1) While listed here as a static muscle, the rectus abdominis actually becomes active during the shrimping action.

2) Your arm's angle relative to your chest as you push an opponent away dictates the part of the pectoral muscle that's used. The lower pectorals are the strongest, and they're triggered when you push your arms down toward your hips. Using the bridge to push your hips off the ground (as shown) helps get the strongest pushing angle.

Plow
Stretches shoulders and spine

Supine twist
Improves spine flexibility

SINGLE-LEG TAKEDOWN

This takedown is often taught as a way for a person on the ground to knock down a standing opponent. However, it can also be used from a standing position. This technique is dangerous to the opponent because it attacks the outside of the knee and pushes it sideways toward the centerline, risking serious injury to the knee. By coming in from the side, this technique is more protected from counterattacks to the face than some other takedowns.

Speed (8 of 10)

Speed is essential since it's easy to counterattack or retreat from this technique. Being closer to your opponent (e.g., when you're already on the ground, at the feet of your opponent) makes this technique easier to execute. Performing this takedown from a standing position usually requires a fake before diving in; the standing entry is very dangerous and should be practiced with great care.

Power (6 of 10)

Power is generated from the thrust of the opposite or outside leg and the extension of the body. This drive extends through the body and into the outside of the opposite shoulder.

Accuracy (8 of 10)

You need to push your opponent's front knee (the one bearing more weight) inward from the outside. The drive should also move downward so that his leg cannot be pulled away easily. The impact point on the leg should be at, or slightly below, the outside edge of the knee. While this move is also taught as an attack to the front of the knee, that angle requires a great deal more force and is thus harder to perform successfully.

KEY EXERCISES

Leap frog + crawl (page 129)
Improves lower-body explosive power and upper-body strength; improves agility

Mountain climber
Improves lower-body power

Seated band row
Strengthens traps, lats, and delts

Key Dynamic Muscles

Arm movement: biceps (unseen), posterior deltoid, latissimus dorsi, pectorals (unseen)

Body drive: gluteus maximus, quadriceps (unseen), calves

Key Static Muscles

Trapezius, deltoids, rectus abdominis (unseen)

Primary Kinetic Chains

Posterior, leg extension

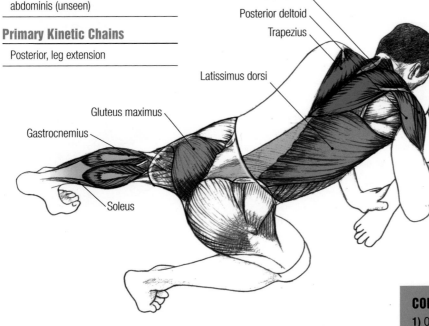

Middle deltoid

Posterior deltoid

Trapezius

Latissimus dorsi

Gluteus maximus

Gastrocnemius

Soleus

Kneeling lunge
Stretches hip flexors and quads

High-elbow shoulder stretch
Stretches shoulders and triceps

COMMENTS

1) One of the more difficult aspects of this technique is getting the outside of your shoulder (right shoulder, as illustrated) onto the outside of your opponent's knee without getting kicked in the face. This is accomplished by placing your outside (left) hand on his foot, your right hand behind his leg, and guiding your shoulder into place. Be careful not to let your shoulder slip from your opponent's knee to the base of your neck as the neck or collarbone can be easily injured.

DOUBLE-LEG TAKEDOWN

An effective way of taking down a standing opponent, this move is dangerous to both you and your partner. As you enter the technique, you risk receiving a counterstrike to anywhere from your face to your groin. Your opponent, on the other hand, risks being picked up and driven into the ground. Since this technique comes in from the side, it's more protected from counterattacks to the face than other takedowns are.

Speed (8 of 10)

Speed is essential since it's easy to counterattack or retreat from this technique. Executing this takedown requires either the element of surprise or a fake before committing to the move.

Power (6 of 10)

Power is generated from the forward thrust of both legs and the extension of the body. This drive extends through the body and into the lead shoulder. The line of power into your opponent usually leads to one of three finishes: Driving upward may lift the opponent off the ground and dump him on his back as you stand upright; driving downward takes the opponent to the ground and you continue using your forward momentum to roll by (or over) your opponent; driving straight forward uses your added weight to slam your opponent into the ground. This last position can be effective but is exceedingly dangerous and should be practiced with extreme care.

Accuracy (8 of 10)

The impact of your shoulder into your opponent's abdomen must coincide closely with your reaching around and grabbing his legs. While the pressure on your opponent's legs won't stop him from stepping back, it will impede him, which makes the takedown possible. Since a common defense from this technique is to splay the legs outward and back as you pitch forward and drive the attacker into the floor, you need to catch his legs early and pull them toward you to prevent this.

KEY EXERCISES

Burpie
Improves full-body explosive power

Mountain climber
Improves lower-body power

Seated band row
Strengthens traps, lats, and delts

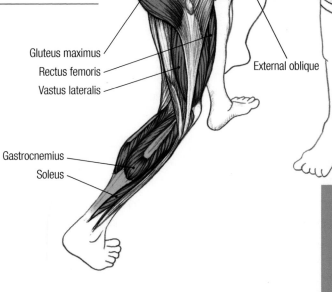

Labels on diagram:
- Trapezius
- Latissimus dorsi
- Teres major
- Deltoid
- Gluteus maximus
- Rectus femoris
- Vastus lateralis
- External oblique
- Gastrocnemius
- Soleus

Key Dynamic Muscles

Arm movement: deltoids, biceps (unseen), latissimus dorsi, teres major, pectorals (unseen), trapezius

Body drive: gluteus maximus, quadriceps, calves

Key Static Muscles

Abdominals

Primary Kinetic Chains

Posterior, leg extension

Deadlift (page 128)
Strengthens glutes, quads, and traps

Kneeling lunge
Stretches hip flexors and quads

COMMENTS

1) Be careful not to hit your opponent with your shoulder near the base of your neck, as the neck or collarbone can be easily injured.

2) This technique has several controversial variations. One involves spearing your opponent in the stomach with the top of your head; this can cause serious neck injuries and should be avoided. Another places your front leg to the outside of your opponent's leg, which gives the move a bit more stability and power but eliminates protection against a kick to the groin or face (unlike stepping in between the legs, as illustrated).

ROLLS & FALLS

Rolls and falls are essential to protect the body in any martial art that involves throws and techniques that might involve going to the ground. For example, if you're practicing wrist techniques, even if a throw isn't planned, sometimes the technique can be applied with so much vigor that you may very well be unexpectedly forced to the ground.

Falls have two general categories: hard and soft. Hard falls require a slapping of the hands or feet on the ground, which distributes the fall's energy so that the body and internal organs don't take the brunt of the blow. The soft fall (sometimes called a soft roll) requires no slap; it's smooth and blends with the force of the impact as you go to the ground.

Learning to roll and fall from a variety of positions and situations is very important. It has been said with more than just a little truth that if you have to think when you fall, your technique will be too late. In other words, situations where you have to either fall or roll arise quickly and unexpectedly, and you must be able to reflexively do the right thing to protect your body. This includes such basic concepts as keeping your head tucked and out of danger of hitting the ground, learning to *kihap* (yell) on impact, and not jamming or breaking an arm or shoulder in a vain attempt to catch yourself as you tumble.

Rolls & Falls

- Forward Roll
- Backward Roll
- Back Fall
- Side Fall
- Face Fall
- Air Fall

FORWARD ROLL

The forward roll is one of the most basic moves in martial arts that involve falling.

Speed (2 of 10)

Speed is usually determined by the cause of the forward roll and is thus largely defined by your momentum. Being pushed forward generally generates more momentum (and thus speed) than if you simply tripped; however, this won't cause a significant change in the overall technique.

Power (2 of 10)

Maintaining a rounded body position during the roll requires a small but varying amount of power, but this is sometimes difficult, such as when you're thrown forward and down. In this case, you may have to absorb a substantial blow to your back as you try to blend into the roll. Many schools teach students to stand up at the end of a roll, but if you have too much momentum, you may need to execute a second roll before standing. If you have too little momentum, you may need a hard tuck of your lower leg and a forward lunge with your body to stand up.

Accuracy (6 of 10)

Maintaining a smooth, rounded contour of your legs, body, and arms is essential in avoiding injury during a roll, especially when rolling on a hard surface. This roundness requires good static-muscle tension throughout the body. It's equally important to keep your head and neck protected, which you can achieve with a well-coordinated chin tuck and slight head twist.

KEY EXERCISES

T + opposite toe touch (page 129)
Develops balance; strengthens legs and core

Dip
Strengthens triceps

Handstand push-up (page 128)
Enhances balance and core and upper body strength

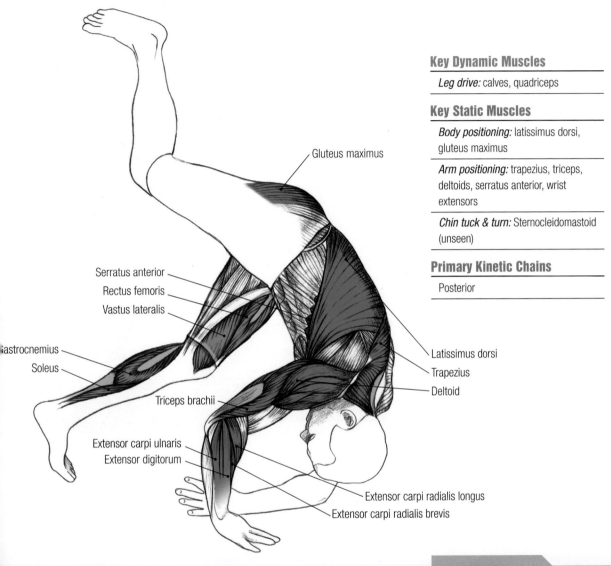

Gluteus maximus

Serratus anterior
Rectus femoris
Vastus lateralis

Gastrocnemius
Soleus

Triceps brachii

Extensor carpi ulnaris
Extensor digitorum

Latissimus dorsi
Trapezius
Deltoid

Extensor carpi radialis longus
Extensor carpi radialis brevis

Key Dynamic Muscles

Leg drive: calves, quadriceps

Key Static Muscles

Body positioning: latissimus dorsi, gluteus maximus

Arm positioning: trapezius, triceps, deltoids, serratus anterior, wrist extensors

Chin tuck & turn: Sternocleidomastoid (unseen)

Primary Kinetic Chains

Posterior

Roll around (page 129)
Warms up back and hips

Plow
Stretches shoulders and spine

COMMENTS

1) One of the biggest beginner problems in the forward roll occurs when your arms collapse when you first apply weight to them. This results in a dramatic loss of the smooth, rolling contour, and other parts of the body (usually the head or shoulder) may receive a significant blow.

BACKWARD ROLL

This roll is often taught in conjunction with the forward roll (page 106) because they share many similarities.

Speed (2 of 10)

Speed is usually determined by the cause of the backward roll and is thus largely defined by your momentum. Being pushed backward generally generates more momentum (and thus speed) than if you simply tripped; this won't cause a significant change in the overall technique.

Power (2 of 10)

Maintaining a rounded body position during the roll requires a small but varying amount of power, but this is sometimes difficult, such as when you perform an axe kick that's caught high and you're thrown backward. In this case, you may have to absorb a substantial blow to your back as you try to blend into the backward roll. Many schools teach students to stand up at the end of a roll, but if you have too much momentum, you may need to execute a second roll before standing. If you have too little momentum, you may need to sharply push with your arms, primarily using your triceps and deltoids, to stand up.

Accuracy (6 of 10)

Maintaining a smooth, rounded contour of your legs, body, and arms is essential in avoiding injury during a roll, especially when rolling on a hard surface. This roundness requires good static-muscle tension throughout your body. It's equally important to keep your head and neck protected, which you can achieve with a well-coordinated chin tuck and slight head twist.

KEY EXERCISES

Military press
Strengthens delts, pecs, and triceps

Warrior 1
Strengthens lower body; stretches quads and shoulders

Neck rotation/stretch
Improves neck mobility

Key Dynamic Muscles

Arm drive: deltoids

Body twist and rounding: rectus abdominis, obliques (unseen), hip flexors, sartorius

Key Static Muscles

Body rounding: quadriceps

Chin tuck & turn: sternocleidomastoid

Primary Kinetic Chains

None

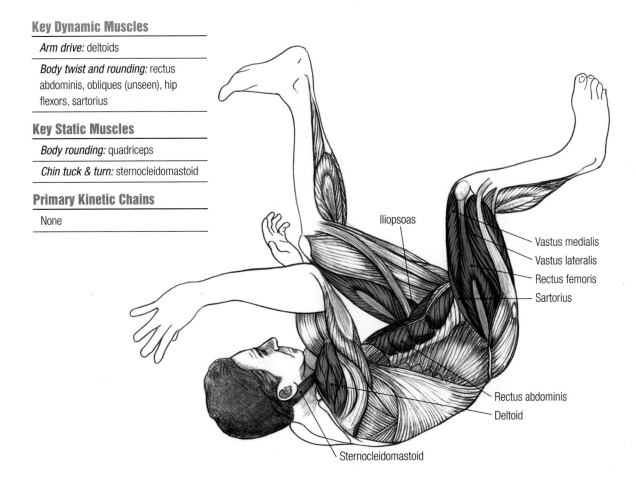

Iliopsoas

Vastus medialis

Vastus lateralis

Rectus femoris

Sartorius

Rectus abdominis

Deltoid

Sternocleidomastoid

Roll around (page 129)
Warms up back and hips

Plow
Stretches shoulders and spine

COMMENTS

1) While forward and backward rolls have many similarities, they differ greatly in how you stand at the end of the technique. The backward roll, especially when done slowly, requires a powerful thrust with the arms to stand up, while the forward roll uses a thrust from the legs.

BACK FALL

The back fall, along with the side fall (page 112), is one of the two most basic falling techniques in martial arts. This fall uses the simultaneous slapping of both arms plus, to a much lesser extent, leg extension to redistribute the fall's energy onto the back—away from vital organs and fragile body parts.

Speed (8 of 10)
Speed is most important when you slap the ground with your hands at the moment of impact. In general, a faster slap absorbs more energy, which results in a more efficient fall.

Power (8 of 10)
The speed and power of the slapping arms directly affects the amount of energy that's redirected away from the vital organs—more power in the slap means more protection for your organs. The classic back fall calls for both hands slapping at the same time.

Accuracy (8 of 10)
The timing of the slap is critical in diverting energy away from the vital organs during a fall. If the slap is too late, your body will have already taken the impact of the fall. If the slap is too early, it's much less efficient at absorbing the fall's energy. While slapping too early is better than slapping too late, it's much better overall to slap at the moment of impact.

KEY EXERCISES

One-arm dumbbell row
Strengthens traps

Crunch (feet up)
Strengthens core muscles

Cat/cow stretch
Stretches back, chest, and neck

Key Dynamic Muscles

Arm cock (not pictured): pectorals, anterior deltoid

Arm slap: trapezius, posterior deltoid, triceps, brachioradialis, pronators (unseen)

Body flexion: rectus abdominis

Leg extension: quadriceps

Key Static Muscles

Pectorals, deltoids

Primary Kinetic Chains

Posterior, arm extension

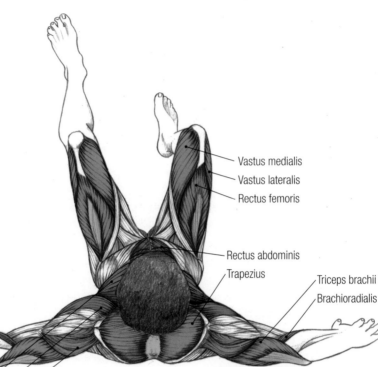

Vastus medialis
Vastus lateralis
Rectus femoris
Rectus abdominis
Trapezius
Triceps brachii
Brachioradialis
Pectoralis major
Anterior deltoid
Posterior deltoid

COMMENTS

1) The single most important thing to remember with these types of falls is to keep your chin tucked to prevent your head from hitting the ground or your neck from suffering whiplash.

2) A *kihap/kiai* (yell) is usually taught with hard falling techniques to help coordinate the tensioning of your body's muscles, let air from your lungs, and tighten your torso muscles, all of which serve to reduce the jostling of your internal organs.

3) At the moment of impact, your hips and legs should be above the plane of the floor to protect your hips; bending at the waist extends the moment of impact so that the energy of the fall has a longer time to dissipate.

Locust
Stretches and strengthens the back body

Arm-across-chest stretch
Stretches shoulders

SIDE FALL

The side fall, along with the back fall (page 110), is one of the two most basic falling techniques in martial arts that involve falling. This fall uses the orientation of the body and the slapping of the bottom arm and leg to redistribute the fall's energy onto the side—away from vital organs and fragile body parts.

Speed (6 of 10)

Speed is most important when you slap the ground with your leg and/or arm at the moment of impact. In general, a faster slap absorbs more energy, which results in a more efficient fall.

Power (9 of 10)

The speed and power of the leg and/or arm slap directly affects the amount of energy that's redirected away from the vital organs—more power in the slap means more protection for your organs. The classic side fall (or break fall, as it's sometimes referred to) calls for both your bottom arm and leg to slap at the same time, but sometimes you're only able to slap with one of them. For example, there are times when the position of your body only allows the lower arm to slap. This arm-only slap is more dangerous but at times is necessary.

Accuracy (6 of 10)

The timing of the slap is critical in diverting energy away from the vital organs during a fall. If the slap is too late, your body will have already taken the impact of the fall. If the slap is too early, it's much less efficient at absorbing the fall's energy. While slapping too early is better than slapping too late, it's much better overall to slap at the moment of impact.

KEY EXERCISES

One-arm dumbbell row
Strengthens traps

Band leg adduction
Strengthens adductors

Side crunch
Strengthens obliques

Key Dynamic Muscles

Arm cock (not pictured): pectorals, anterior deltoids

Leg slap: gluteus medius, obliques

Arm slap: trapezius, posterior deltoid, triceps, pronators (unseen)

Key Static Muscles

Rectus abdominis, middle deltoid, quadriceps, sartorius

Primary Kinetic Chains

Lateral, arm extension

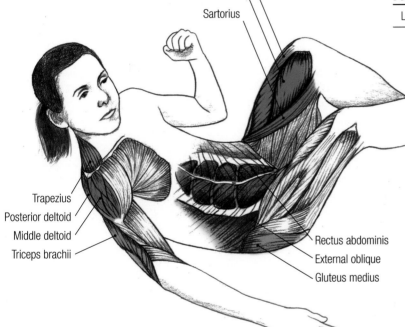

Rectus femoris

Vastus medialis

Sartorius

Trapezius

Posterior deltoid

Middle deltoid

Triceps brachii

Rectus abdominis

External oblique

Gluteus medius

COMMENTS

1) The single most important thing to remember with these types of falls is to keep your chin tucked to prevent your head from hitting the ground or your neck from suffering whiplash.

2) While the ball of the foot of the top leg is sometimes used to slap, the energy that it can distribute is minimal compared to the bottom arm and leg. However, the positioning of your top leg is important to prevent the insides of your knees from smashing together.

3) At the moment of impact, your hips and legs should be above the plane of the floor to protect your hips; bending at the waist extends the moment of impact so the fall's energy has a longer time to dissipate.

Arm-across-chest stretch
Stretches shoulders

Pigeon
Stretches hips, quads, and groin

FACE FALL

This important break fall protects you when you land on your front. The face fall can be done two ways: from a forward fall or with a jump.

Speed (4 of 10)

Speed in the face fall is more regulated than maximized. Speed must be used to synchronize the arm slap (and leg slap, when jumping) with the moment of impact. Both the jump and the landing also require careful timing.

Power (7 of 10)

Jumping Takeoff: The takeoff involves two actions: 1) an upward leg thrust that propels your body upward and forward; and 2) pulling back the arms in preparation for the slap. You have to balance these two actions so you can land flat on your front side. **Landing**: The landing requires a hard, simultaneous slap with the forearms (and the balls of both feet, when performed with a jump). This action must be done in conjunction with your body landing in a piked or bridge position, where your hands and feet are close enough together that your butt is deliberately positioned well off the ground, protecting your pelvis from impact.

Accuracy (8 of 10)

Jumping Takeoff: The jump's timing requires careful coordination of both the jump upward and the rotation forward. **Landing**: Simultaneously delivering sharp, hard blows with all four limbs is critical for protecting your vital organs during a fall. If the slap is too late, your body will have already taken the impact of the fall. If the slap is too early, it will be less efficient at absorbing the fall's energy. While slapping too early is better than slapping too late, it's much better overall to slap at the moment of impact.

KEY EXERCISES

Clapping push-up (page 128)
Improves upper-body explosive power

Dumbbell fly
Strengthens pecs

Plank
Strengthens core and deltoids

Key Dynamic Muscles—Jump (not pictured)

Leg extension: quadriceps, calves

Body pike: rectus abdominis

Arm preparation: trapezius

Key Static Muscles

None

Primary Kinetic Chains

Posterior, leg extension

Key Dynamic Muscles—Landing

Arm slap: pectorals, anterior deltoid (unseen), triceps, pronators (unseen)

Foot slap (from jumping version; not pictured): quadriceps, ankle dorsiflexors

Key Static Muscles

Rectus abdominis, trapezius, middle deltoid; *(from falling version)* gluteus maximus, quadriceps, calves

Primary Kinetic Chains

Arm extension

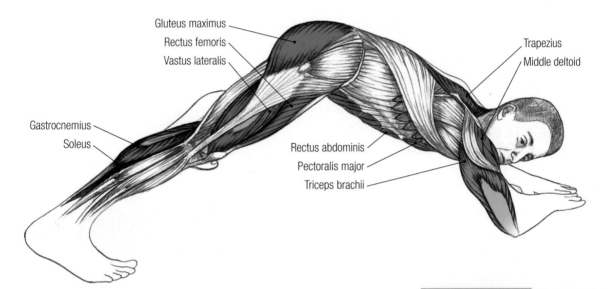

Gluteus maximus
Rectus femoris
Vastus lateralis
Gastrocnemius
Soleus
Rectus abdominis
Pectoralis major
Triceps brachii
Trapezius
Middle deltoid

Downward-facing dog
Strengthens arms and legs; stretches shoulders, back, and hamstrings

High-elbow shoulder stretch
Stretches shoulders and triceps

COMMENTS

1) The anterior kinetic chain consists of muscles along the front of the body, and it includes the quadriceps in the legs all the way up to the pectoral muscles in the chest. Proper tension of this kinetic chain is critical in executing this fall.

AIR FALL

The air fall, when done as an exercise, combines a jump with a three-quarter flip and a landing with a side fall. This is an important fall to learn because it's used to land after a large number of techniques ranging from wrist throws to cane throws. This technique is broken up into two parts: the takeoff and the landing.

Speed (7 of 10)

Speed in the air fall is more regulated than maximized. **Takeoff**: The jump requires that its height and rotational velocity be timed so that your body lands after three-quarters of a flip, precisely on the side. **Landing**: Speed must be used to synchronize the arm and leg slap with the moment of impact.

Power (5 of 10)

Takeoff: The takeoff requires two actions: 1) an upward thrust with the supporting leg, which propels the body upward; and 2) a hard heel kick with the upper leg, which gives your body the forward rotation. You need to balance these two actions so you can land flat on your side. **Landing**: The landing is a side fall that requires a hard, simultaneous slap with both the lower arm and lower leg. The only real distinction from a side fall is that the side fall can be performed from a backward or forward rotation, but an air fall always uses a forward rotation.

Accuracy (8 of 10)

Takeoff: The timing of the jump requires careful coordination of the upward jump and the back heel kick for rotation. **Landing**: The timing of the slap is critical for diverting energy away from the vital organs during a fall. If the slap is too late, the body will have already taken the impact of the fall. If the slap is too early, it will be less efficient at absorbing the fall's energy. While slapping too early is better than slapping too late, it's much better overall to slap at the moment of impact.

KEY EXERCISES

Burpie
Improves full-body explosive power

One-leg hop
Improves lower-body explosive power

One-arm dumbbell row
Strengthens trapezius

Gluteus maximus
Biceps femoris
Gastrocnemius
Soleus

Key Dynamic Muscles—Jump:

Bottom-leg extension: quadriceps, calves

Top-leg heel kick: gluteus maximus, hamstrings

Body tuck: rectus abdominis

Primary Kinetic Chains

Posterior, leg extension

Key Dynamic Muscles—Landing:

Leg slap: gluteus medius (unseen), vastus lateralis

Leg splay: sartorius

Arm cocking (not pictured): pectorals, deltoids

Arm slap: deltoids, triceps, pronators (unseen)

Key Static Muscles

Abdominals, quadriceps

Primary Kinetic Chains

Lateral, arm extension

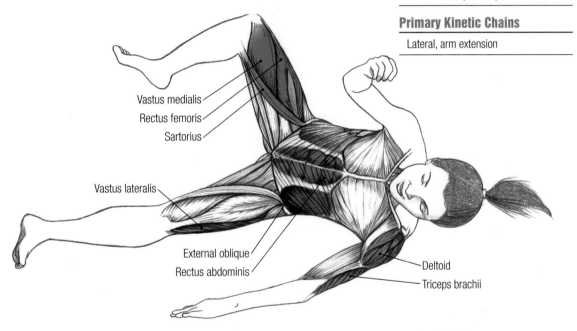

Vastus medialis
Rectus femoris
Sartorius
Vastus lateralis
External oblique
Rectus abdominis
Deltoid
Triceps brachii

Arm-across-chest stretch
Stretches shoulders

Seated twist
Improves spine flexibility

COMMENTS

1) The sartorius is designated a dynamic muscle in this fall because it's used to pull the top knee upward and away from the lower knee so they don't clash upon landing.

WEAPONS

The use of weapons in martial arts is highly varied. Some arts such as kendo are exclusively dedicated to a specific weapon. Other arts might concentrate on defenses against attacks from an assortment of weapons. Finally, some arts teach you how to utilize a number of weapons both in defense and offense. Regardless of how a weapon is used, it can greatly magnify your reach, speed, and power.

This section highlights four weapons: the shinai (bamboo sword), cane, tonfa, and short stick. All these weapons are used to hit an opponent, but they do so in a variety of different ways. With the shinai, virtually the entire body is involved in the strike, while the other three weapons focus on striking with a subset of the body. These weapons can also be used to defend against an attack.

Weapons

- **Shinai Strike**
- **Cane Thrust**
- **Tonfa Thrust**
- **Short Stick Strike**

SHINAI STRIKE

One of the single most refined moves in martial arts, the classic kendo strike to the head with a shinai, or bamboo sword, requires an exacting combination of speed, power, and accuracy. This technique perfectly illustrates the concept of relaxing muscles so they can be driven forward, then moving them dynamically so they can accelerate the body, and finally tensing them so they can drive weight into the strike.

Speed (7 of 10)

Speed is generated from a whiplike motion that starts with the drive of the back foot and ends in unison with the forward wrist snap. The coordination of the footwork, arm swing, and extension, combined with the ultimate wrist snap, determines the blow's final speed. The intricate coordination of chest, back, and shoulder muscles to swing the arm forward is very complex; people have studied it for years trying to describe all facets of the movement.

Power (8 of 10)

Arm extension, wrist snap, and footwork ultimately deliver the crushing power in this strike. Some key factors are:

Hand pronation at impact: At the moment of impact, the sword's handle tends to kick upward, so your hands must be turned down at the moment of impact so that they firmly cover the top of the shinai. This prevents loss of much of the blow's power.

Arm swing: Beginners tend to pull their arms in toward their body in a vain attempt to generate more power into the blow. Real power comes from whipping out the sword, which requires arm extension at impact.

Accuracy (10 of 10)

The target of this blow is the top of your opponent's rounded helmet, which is very hard to hit. Only a perfectly placed strike will allow power to extend into your opponent. However, it's not uncommon to hear of people being hit so hard in the helmet by a well-placed blow that they're driven down, bruising their heels.

KEY EXERCISES

Warrior 1
Strengthens lower body; stretches quads and shoulders

Barbell/dumbbell pullover
Strengthens pecs, triceps, and lats

High-elbow shoulder stretch
Stretches shoulders and triceps

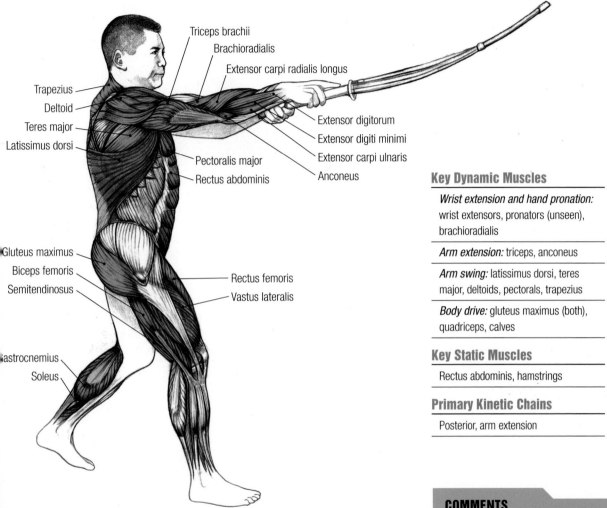

Triceps brachii
Brachioradialis
Extensor carpi radialis longus
Trapezius
Deltoid
Teres major
Latissimus dorsi
Pectoralis major
Rectus abdominis
Extensor digitorum
Extensor digiti minimi
Extensor carpi ulnaris
Anconeus
Gluteus maximus
Biceps femoris
Semitendinosus
Rectus femoris
Vastus lateralis
Gastrocnemius
Soleus

Key Dynamic Muscles

Wrist extension and hand pronation: wrist extensors, pronators (unseen), brachioradialis

Arm extension: triceps, anconeus

Arm swing: latissimus dorsi, teres major, deltoids, pectorals, trapezius

Body drive: gluteus maximus (both), quadriceps, calves

Key Static Muscles

Rectus abdominis, hamstrings

Primary Kinetic Chains

Posterior, arm extension

COMMENTS

1) In addition to adding power to the blow, hand pronation at impact is also necessary to protect your thumbs from the shinai's back kick during impact. Beginners who forget to pronate their hands may experience severe thumb damage.

2) The kendoist strives to deliver a strike with "*ki, ken, tai, ichi,*"— meaning "spirit, sword, body are one."

Wide-leg forward bend + shoulder stretch
Stretches hamstrings and shoulders

Kneeling forearm stretch
Stretches wrists and forearms

CANE THRUST

The cane thrust is not very strong or fast compared to many other cane techniques, but because it strikes with such a small surface area (the butt of the cane), it's hard to block and can inflict substantial local damage. Targets for this strike are usually the stomach, face, or throat.

Speed (5 of 10)

The arm's forward thrust generates most of the speed. It's the final segment of the usual drive chain starting from the back foot, and it requires the cane to be more or less parallel to the floor and pushed in a straight line. If the cane is thrust up in a curve, it tends to skip off the intended target.

Power (4 of 10)

The forward thrust of the arm is responsible for the strike's power, but the power is difficult to deliver. Key factors in power generation include:

Forearm alignment: The shaft of the cane must line up precisely with your forearm at the moment of impact. When this fails, your wrist bends and the strike becomes mostly ineffective.

Arm supination: The upward twist of the forearm stiffens the forearm at impact. While both supination and pronation are taught with this strike, supination is the most common and strongest.

Hip lock: When striking someone in the stomach, sliding your striking elbow in front of your striking hip lends support to the blow and negates the rebound that usually occurs from hitting someone near their center of mass.

Palm on head of cane: Resting your striking hand's palm on the cane's curve while still keeping it in line with the shaft allows the line of force moving down your forearm to extend straight down the cane and into the target.

Accuracy (8 of 10)

This strike's accuracy requirements are unusually high because targets are limited. Striking to the stomach is the easiest, while striking to the face and throat are substantially harder since the targets are small and demand great precision to hit consistently.

KEY EXERCISES

Warrior 1
Strengthens lower body; stretches quads and shoulders

Lunge + twist
Enhances hip flexibility while developing core power

Fingertip push-up (page 128)
Strengthens hands, wrists, pecs, and triceps

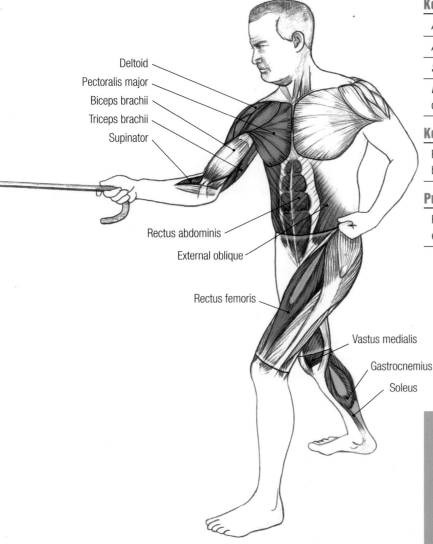

Deltoid
Pectoralis major
Biceps brachii
Triceps brachii
Supinator
Rectus abdominis
External oblique
Rectus femoris
Vastus medialis
Gastrocnemius
Soleus

Key Dynamic Muscles

Arm supination: supinator

Arm thrust: deltoids, pectorals, triceps

Shoulder twist: obliques

Body drive: gluteus maximus (unseen), quadriceps, calves

Key Static Muscles

Rectus abdominis, obliques, quadriceps, biceps

Primary Kinetic Chains

Posterior, hip turn, shoulder turn, arm extension

Wide-leg forward bend + shoulder stretch
Stretches hamstrings and shoulders

Kneeling forearm stretch
Stretches wrists and forearms

COMMENTS

1) Cane thrust attacks come in many varieties, and the two common versions pictured here have major trade-offs in the range, speed, power, and stability of the weapon. In the illustration above, the attacking elbow is extended far from the body. This enhances the range and speed of the technique but reduces power and stability. The picture on the facing page shows the attacking elbow locked into the hip, which increases the power and stability of the strike but diminishes the range and speed of the blow. This kind of trade-off is common in martial arts techniques.

TONFA THRUST

The tonfa thrust basically blends a reverse punch and a palm heel strike. The illustrated strike uses the butt of the tonfa as the striking surface and also involves a second tonfa, which augments a forearm block as the first tonfa is used to strike out. Tonfa are also used as swinging weapons.

Speed (5 of 10)

The speed is similar to that of a palm heel strike. Keeping the arm muscles relaxed by not grabbing the tonfa too tightly allows faster speeds. Since tonfa are used for both thrusts and swings, both linear speed (for thrusts) and side-to-side speed (for slashing or slapping out with the sides of the weapon) are very important.

Power (6 of 10)

The tonfa's final drive is based on the thrust of the palm heel. A key factor in power generation includes:

Arm pronation/supination: This technique requires either arm pronation or supination (both stiffen the forearm) as the tonfa strikes. In general, pronation is used more with straight or downward thrusts, while supination is used more with rising thrusts.

Accuracy (6 of 10)

The point of contact, especially when the body is the target, requires less precision than some other strikes. Many teach that a thrusting strike anywhere between the solar plexus and the hip (the mid-section) is a reasonable target. Strikes to other targets, such as the face and limbs, require substantially more accuracy, which is why blows such as the slapping action are used more commonly against those targets.

KEY EXERCISES

Warrior 1
Strengthens lower body; stretches quads and shoulders

Lunge + twist
Enhances hip flexibility while developing core power

Push-up
Strengthens pecs, triceps, and wrist extensors

Key Dynamic Muscles

Arm pronation or supination: pronators (unseen); supinator

Arm extension: deltoids, trapezius, triceps

Body twist: obliques (unseen)

Body drive: gluteus maximus, quadriceps, calves

Key Static Muscles

Rectus abdominis, quadriceps, calves, trapezius

Primary Kinetic Chains

Posterior, hip turn, shoulder turn, arm extension

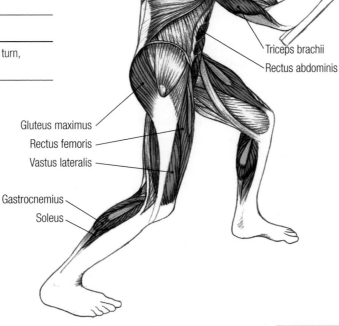

Supinator

Trapezius

Deltoid

Triceps brachii

Rectus abdominis

Gluteus maximus

Rectus femoris

Vastus lateralis

Gastrocnemius

Soleus

Wide-leg forward bend + shoulder stretch
Stretches hamstrings and shoulders

Kneeling forearm stretch
Stretches wrists and forearms

COMMENTS

1) A second tonfa is used to protect the striker, but the striking tonfa, whose main body lies down the striking forearm, also adds protection, especially against an armed opponent.

SHORT STICK STRIKE

The so-called short stick is a broadly defined weapon that ranges from six inches to over two feet in length and is made of everything from light, whipping material to heavy, rigid material. These different sticks necessitate a trade-off between speed and power. Common "hard" targets are the head, forearms (shown here), and shins, and the joints such as the elbows and knees; "soft" targets include the groin, abdomen, and kidneys.

Speed (6 of 10)

Speed is highly dependent on the length and heaviness of the stick, as well as the final flick of the wrist (sometimes called a "drummer's flick"). Grip is also very important, and most of the grip's strength is between the thumb and index finger, which make a pivot point. Holding the rest of the fingers lightly allows the weapon to snap.

Power (5 of 10)

There's a trade-off between the speed and power of the blow. In general, pronating the wrist adds more power—twisting the bones in the forearm transfers the weight of the body mass more efficiently—but reduces speed. Depending on the blow's direction (e.g., outside to inside, inside to outside, or straight down), the hip turn, shoulder twist, and arm extension will play a larger or smaller role in power generation. An interesting practice drill is to have one person hold a padded stick and to have a partner grab them over their arms. The grabbed person should then try to strike with the stick. The overarm grab (depending on whether it's over the upper or lower arms) largely prevents the use of the hips and/or shoulder, thus limiting the blow's power to the arm extension and the wrist snap.

Accuracy (8 of 10)

While the common hard targets listed above take advantage of the short stick's hardness, other targets are also possible, such as the groin (which usually requires an uppercut) and the floating or lower ribs. It's important to be able to accurately place "stopping" blows (usually to the legs or head), which stop the attack immediately. However, if an opponent attacks with a weapon, then his weapon should be considered the highest priority. In this case, striking your opponent's forearm (the radius) on the thumb side is often most effective because it can numb or even break the arm.

KEY EXERCISES

Warrior 1
Strengthens lower body; stretches quads and shoulders

Lunge + twist
Enhances hip flexibility while developing core power

Woodchopper
(page 129)
Strengthens obliques and shoulders

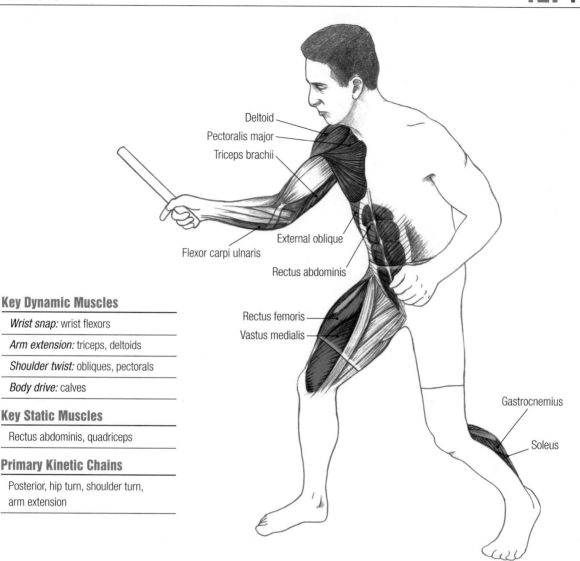

Deltoid
Pectoralis major
Triceps brachii
External oblique
Flexor carpi ulnaris
Rectus abdominis
Rectus femoris
Vastus medialis
Gastrocnemius
Soleus

Key Dynamic Muscles

Wrist snap: wrist flexors

Arm extension: triceps, deltoids

Shoulder twist: obliques, pectorals

Body drive: calves

Key Static Muscles

Rectus abdominis, quadriceps

Primary Kinetic Chains

Posterior, hip turn, shoulder turn, arm extension

Wide-leg forward bend + shoulder stretch
Stretches hamstrings and shoulders

Kneeling forearm stretch
Stretches wrists and forearms

COMMENTS

1) Before hitting targets with a stick, make sure that the stick has no breaks or splinters in it. Significant injuries, especially to the eyes, can occur when a stick breaks.

2) A common alternative grip is with the ring and pinky fingers gripping and acting as a pivot point. This changes the wrist snap and thus the speed and power of the strike.

APPENDIX 1: INSTRUCTIONS FOR MARTIAL ARTS–SPECIFIC EXERCISES

Instructions for the less-traditional exercises featured in this book.

BODY DRAG—PULL

Partner lies on back with legs bent, feet off the ground, and arms crossing chest. Sit behind your partner with your legs on either side of his hips and grip under his triceps/armpits. Bend your knees then powerfully straighten them to launch yourself backward from your partner. Use your back muscles to pull your partner to you. To prevent injuries, especially when working with heavier people, do not start the pull with a rounded back.

CLAPPING PUSH-UP

Perform a standard push-up but explode off the ground so that you have enough time to clap your hands; return your palms to the ground before your body lands, turning your head slightly to one side in case the catch is late and your face hits the floor. Practice first from the knees, keeping your body and hips straight. Add a double clap for an extra challenge.

CROSS-BODY DOWNWARD BAND PULL

Grip the band by your opposite ear. Pull down and across your chest to your hip. Slowly reverse direction.

DEADLIFT

Partner lies on side. Stand with your shins against your partner's back and butt. Squat down with a straight back and grip your partner's uniform at roughly the near-side shoulder and the knee of her lower leg. Keeping your back straight and your partner against your shin, lift your partner by straightening your legs.

FINGERTIP PUSH-UP

Perform a push-up with only your fingertips on the floor; this requires greater wrist stability. Using fewer fingertips increases the pressure on the finger joints; thumb-only push-ups must be done carefully because they can stress the thumb joints greatly.

HALF MOON + CRUNCH

Assume half moon (against a wall if balance is an issue). Keeping the standing leg and hip still, perform a side crunch until your upper body is above hip level.

HANDSTAND PUSH-UP

Assume a handstand (partners may be required to hold your legs if you don't have adequate upper-body strength). Bend your elbows to touch your head to the floor, then push back up.

HIGH SHOOT

1. Lie on your back. **2.** Roll up to a seated position and turn to one side (right shown). **3.** Holding your body off the ground, turn 90 degrees and place your hands on the ground. **4.** Shoot your left leg under your body, and turn another 90 degrees in the same direction. **5.** Place your hands on the floor with your chest up toward the ceiling. **6.** Turn another 90 degrees and again turn over into a push-up position. **7.** Shoot your left leg under your body. **8.** Turn another 90 degrees in the same direction, and place your hands on floor with your chest up toward the ceiling. **9.** Sit and roll back. You've now rotated your body 360 degrees from where you started and turned 720 degrees in relationship to your spine.

INCHWORM

Start in a forward bend. Keeping your legs straight throughout the exercise, walk your hands out until you're in a push-up position. Do a push-up, then walk your legs back in. Continue moving forward.

IN-TO-OUT BAND PULL

Hold a band next to one hip. Grab the band with the opposite hand (keeping your elbow next to your ribs and your palm facing your torso) and, pivoting from your elbow, perform an in-to-out block.

JUMP WITH 180/360-DEGREE TURN

From a standing position, jump and twist 180/360 degrees. Land in a solid stance.

KNEE RAISE

Explosively pull one knee to your chest, alternating legs. Add a hop or jump to make the exercise more dynamic.

LEAP FROG + CRAWL

Partner bends over and protects his head with his hand. Leapfrog over your partner, land, and immediately crawl between your partner's legs.

LOW SHOOT

1. Lie on your back. **2.** Roll up to a seated position and turn to one side (left shown). **3.** Turning 90 degrees, place both forearms on the ground and hold your body off the ground. **4.** Shoot your right leg under your body and turn another 90 degrees in the same direction. **5.** With your chest pointed toward the ceiling again, sit and roll back. You've now rotated your body 180 degrees from where you started and turned 360 degrees in relationship to your spine.

ONE-LEGGED BRIDGE + HIP DIP

Assume a bridge position. Keeping your hips level, extend one leg up to ceiling. Dip your hips to the floor, then return your hips to starting position.

ROLL AROUND

Grab your knees, round your back, tuck your chin, and roll forward, backward, and side to side.

SIDE KICK EXTENSION ALONG WALL

Stand against a wall and chamber your leg for a side kick, keeping your heel on the wall throughout the exercise. Slowly extend and retract kick.

SIT-UP WITH PUNCH

Perform a sit-up, twist your body to one side, and punch; twist your body to the other side, and punch.

SQUAT WITH PARTNER

Perform a squat with a partner draped across your shoulders or hips. Form is very important to avoid injuring the back or knees.

STANDING BAND PULL

Two variations for throwing practice: 1) Stand sideways to a partner, grab one end of the band in each hand, and pull. 2) Stand in front of a partner, grab one end of the band in each hand, and pull.

SUPINE LEG PUSH-DOWN

Lie on your back and hold onto your partner's ankles. Raise your legs to the ceiling. Your partner pushes your legs to the ground, either directly down or to the side. Using your core muscles, prevent your legs from hitting the ground and then bring them back up; do not let your back arch.

T + OPPOSITE TOE TOUCH

From standing, raise one leg behind you and lower your upper body until your body forms a straight line from head to heel. Extend your arms out to the sides in a T. Bend slightly at the waist to touch opposite hand to opposite foot. Return to a T position, take a step forward, and repeat.

T PUSH-UP

Perform a push-up, then open your body to one side, reaching your arm to the ceiling.

TOE WALK

Stand on the balls of both feet, raising your heels as high as possible. Walk.

WARRIOR 2 BAND PULL

Stand in Warrior 2 with a band under your straight-leg foot; grip the other end in your far hand, keeping the closer arm by your side. Starting with the hand at the hip of the straight leg and leading with your elbow, slowly pull the band across your chest until the arm is fully extended. Slowly reverse direction.

WOODCHOPPER

Hold a medicine ball in both hands high to one side. Twist your torso to lower the ball to the other side.

APPENDIX 2: MUSCLES (ALPHABETIZED) & THEIR MOVEMENTS

MUSCLE	MUSCLE ACTION
Adductor brevis	Adducts thigh at hip
Adductor longus	Adducts & medially rotates thigh at hip
Adductor magnus	Adducts thigh
Anconeus	Assists triceps in extending forearm at elbow
Biceps brachii	Flexes forearm at elbow; supinates flexed forearm
Biceps femoris (hamstring)	Flexes leg at knee; extends thigh at hip
Brachialis	Flexes forearm at elbow in all directions
Brachioradialis	Flexes forearm at elbow during midpronation
Deltoid	*Anterior:* flexes & medially rotates arm; *Middle:* abducts arm; *Posterior:* extends & laterally rotates arm
Extensor carpi radialis	Extends & abducts hand at wrist
Extensor carpi ulnaris	Extends & adducts hand at wrist
Extensor digitorum	Extends hand at wrist
Extensor digitorum longus	Dorsiflexes ankle
Extensor hallucis longus	Extends big toe; dorsiflexes ankle
Flexor carpi radialis	Flexes & abducts hand at wrist
Flexor carpi ulnaris	Flexes & adducts hand at wrist
Gastrocnemius (calf)	Plantarflexes ankle; flexes leg at knee
Gluteus maximus	Extends thigh at hip; laterally rotates hip
Gluteus medius	Abducts thigh at hip; medially rotates hip
Gluteus minimus	Abducts thigh at hip, medially rotates hip
Gracilis	Adducts thigh at hip; flexes leg at knee & helps in medial rotation
Iliopsoas	Flexes thigh at hip
Latissimus dorsi	Extends, adducts, & medially rotates upper arm

MUSCLE	MUSCLE ACTION
Obliques, external/internal (abdominals)	Flex & rotate trunk
Obturator externus/internus	Laterally rotates thigh at hip
Pectineus	Adducts & flexes thigh at hip
Pectoralis major	Flexes, adducts, & medially rotates arm
Piriformis	Laterally rotates extended thigh at hip
Pronator quadratus	Pronates forearm
Pronator teres	Pronates forearm; flexes elbow
Quadriceps femoris (quadriceps group)	Extends leg at knee
Rectus abdominus (abdominals)	Flexes trunk
Rectus femoris (quadricep)	Extends leg at knee; flexes thigh at hip
Rhomboids	Retract scapula
Sartorius	Flexes, abducts, & laterally rotates thigh at hip; flexes knee
Semimembranosus (hamstring)	Medially rotates hip
Semitendinosus (hamstring)	Medially rotates hip
Serratus anterior	Elevates/depresses ribs; rotates scapula upward; protracts scapula
Soleus (calf)	Plantarflexes ankle
Sternocleidomastoid	Turns head
Supinator	Supinates forearm
Tensor fascia latae	Abducts, medially rotates, & flexes thigh at hip
Teres major	Extends arm & medially rotates shoulder
Teres minor	Laterally rotates arm
Tibialis anterior	Dorsiflexes ankle
Tibialis posterior	Plantarflexes ankle
Trapezius	Elevates, retracts, upwardly rotates, & depresses scapula
Triceps brachii	Extends forearm at elbow
Vastus intermedius (quadricep)	Extends leg at knee
Vastus lateralis (quadricep)	Extends leg at knee
Vastus medialis (quadricep)	Extends leg at knee

APPENDIX 3: MUSCLE ACTIONS BY JOINT

JOINT	ACTION	MUSCLES			
HIP					
	Flexion	iliopsoas	rectus femoris	sartorius	pectineus	tensor fascia latae
	Extension	hamstrings	gluteus maximus			
	Abduction	gluteus medius	gluteus minimus	sartorius	tensor fascia latae	
	Adduction	adductors (brevis, longus, magnus)	gracilis	pectineus		
	Medial rotation	gluteus medius	gluteus minimus	adductor longus	tensor fascia latae	semimembranosus semitendinosus
	Lateral rotation	obturator externus/internus	piriformis	gluteus maximus	sartorius	
KNEE						
	Flexion	hamstrings	gracilis	sartorius	gastrocnemius	
	Extension	quadriceps femoris (rectus femoris, vastus intermedius, vastus lateralis, vastus medialis)				
	Medial rotation	semitendinosus	semimembranosus	gracilis		
ANKLE						
	Plantarflexion	gastrocnemius	soleus	tibialis posterior	flexor digitorum longus	flexor hallucis longus
	Dorsiflexion	tibialis anterior	extensor digitorum longus	extensor hallucis longus		
SHOULDER						
	Flexion	pectoralis major	anterior deltoid			
	Extension	latissimus dorsi	posterior deltoid	teres major		
	Abduction	middle deltoid				
	Adduction	pectoralis major	latissimus dorsi	coracobrachialis		
	Medial rotation	teres major	pectoralis major	latissimus dorsi	anterior deltoid	
	Lateral rotation	teres minor	posterior deltoid			
	Upward rotation	trapezius	serratus anterior			
	Elevation	trapezius	levator scapulae			
	Depression	trapezius				
	Retraction	trapezius	rhomboids			
	Protraction	serratus anterior				
ELBOW						
	Flexion	brachialis	biceps brachii	brachioradialis	pronator teres	
	Extension	triceps brachii	anconeus			
WRIST						
	Pronation	pronator teres	pronator quadratus			
	Supination	supinator	biceps brachii			
	Flexion	flexor carpi radialis	flexor carpi ulnaris			
	Extension	extensor carpi radialis	extensor carpi ulnaris	extensor digitorum		
	Abduction	flexor carpi radialis	extensor carpi radialis			
	Adduction	flexor carpi ulnaris	extensor carpi ulnaris			
TRUNK						
	Flexion	rectus abdominis	obliques			
	Rotation	obliques				

GLOSSARY

Abs: Common term for "abdominals," the muscles of the abdomen. Includes rectus abdominis, transversus abdominis, and internal and external obliques.

Anterior: To the front of. Opposite of posterior.

Arthritis: Inflammation of a joint. Repeated occurrences will lead to degeneration or permanent damage to the joint.

Calf: Common term for two muscles (gastrocnemius and soleus) located at the back of the lower leg. Both muscles flex the ankle; the gastrocnemius also flexes the leg at the knee.

Clavicle: Long thin bone that connects the sternum to the scapula. Latin for "little key," it rotates like a key on its long axis when the shoulder twists. Also known as the collarbone.

Concussion: A brain injury that causes a change in mental state. A concussion may or may not cause a loss of consciousness.

Delt: Common term for "deltoid," the shoulder muscle that moves the arm.

Dorsal: Latin for "back," dorsal refers to a position or movement to the back of the body. Opposite of ventral.

Dynamic: The characteristic of a movement, usually with force and/or power.

Floating ribs: The bottom two of twelve ribs on each side of the body. While all ribs attach to the spine, only the top ten attach to the sternum or its cartilage. This makes the floating ribs more susceptible to injury.

Grappling: Close, hand-to-hand combat. Wrestling.

Hamstring: Common term for one of the three muscles (semitendinous, semimembranosus, biceps femoris) located at the back of the thigh that extends the hip joint and flexes the knee.

Hyperextension: To open a joint beyond its intended range. This action often leads to injury.

Ki: A complex concept that sometimes is roughly translated as "energy flow" but has many other aspects, including vitality and spirit among many other things. Also spelled *chi* and *qi*.

Kinetic chain: Concept of interconnecting muscles and bones working together, often in complex sequences, to produce strong, effective motion.

Kinetic energy: The energy possessed by a moving object; defined to be equal to one half times the mass of the object times the velocity of the object squared.

Lateral: Latin for "to the side," lateral refers to a position or movement that is away from the spine or central axis. Opposite of medial.

Lat: Common term for "latissimus dorsi," a powerful back muscle.

Medial: Latin for "middle," medial refers to a position or movement that is toward the spine or central axis. Opposite of lateral.

Obliques: Common term for "internal and external obliques," abdominal muscles responsible for torso flexion and rotation.

Pelvis: Bony structure that connects the spine to the legs. From the Latin word for "bowl," it's shaped like a large basin.

Posterior: To the back of. Opposite of anterior.

Pronation: Turning the palm downwards. This twists the two bones of the forearm (the ulna and radius) to make the forearm effectively stiffer and more efficient at transmitting power.

Quadriceps (quads): Common term for the four muscles (rectus femoris, vastus lateralis, vastus medialis, vastus intermedius) located at the front of the thigh. All four muscles extend the leg at the knee, but the rectus femoris also flexes the leg at the hip.

Scapula: Large, flat, triangular shoulder bone that effectively connects the collarbone to the upper arm. From the Greek word "to dig," as the bone looks like a shovel.

Shinai: Bamboo sword used to strike armored opponents in martial arts such as kendo and kumdo.

Shrimping: A groundwork movement in which one strongly twists side to side to escape or change positions relative to an opponent.

Static: Something that is not moving. Note that this does not mean that the position is weak. In fact, many static positions, such as a horseback stance, are very strong.

Sternum: Long, flat bone that connects the collarbone with the first seven ribs. Also known as the breast bone.

Supination: Turning the palm upward. This twists the two bones of the forearm (the ulna and radius) to make the forearm effectively stiffer and more efficient at transmitting power. In general, pronation is more effective in this regard than supination.

Ventral: Latin for "abdomen," this refers to a position or movement to the front of the body. Opposite of dorsal.

INDEX

INDEX

ABOUT THE AUTHORS

Norman Link has been practicing martial arts for over 40 years and is a 7th-degree black belt in yongmudo (formerly known as hapkido). He currently serves on the United States Yongmudo Association board and is the head yongmudo instructor at the University of California Martial Arts Program (www.ucmap.org) in Berkeley, California. He also practices jiujitsu and iaido. In addition to his martial arts training, he has performed extensive study in numerous areas of bioengineering and medical research, ranging from eye muscle stimuli responses to robotic diagnosis of cancers (such as lymphomas) to flash x-ray imaging of bulletproof vests at the moment of impact. He holds a PhD in biomedical/electrical engineering and currently works as a scientist in the Bay Area.

Lily Chou has been practicing martial arts for 15 years and is a 3rd-degree black belt in yongmudo. In addition to dabbling in taekwondo and no-*gi* jiujitsu, she is an editor of health and fitness books, a certified yoga instructor, and the author of *The Martial Artist's Book of Yoga*.

ABOUT THE MODELS

Jon Bertsch has been practicing martial arts for over 30 years and is a 4th-degree black belt in judo. He also practices hapkido and is a programmer in the Bay Area.

David Commins has been practicing martial arts for over 35 years and is a 5th-degree black belt in taekwondo. In addition, he practices yongmudo, judo, and iaido, and is a lawyer in San Francisco.

Luke Commins has been practicing martial arts for over 10 years and is a 1st-degree black belt in taekwondo. He is a member of the UC Berkeley taekwondo demo team.

Kelly Kim has been practicing martial arts for over 15 years and is a 2nd-degree black belt in taekwondo. She is a member of the 2009–2011 U.S. Wushu Team and is currently studying kinesiology and sport biomechanics in the Bay Area.

Susan Link has been practicing martial arts for over 30 years and is a 4th-degree black belt in yongmudo. She has also practiced and competed nationally in taekwondo and is an optometrist in the Bay Area.

Bob Matsueda has been practicing martial arts for over 40 years and is a *renshi* 6th-degree black belt in kendo. He is the head instructor for the Berkeley Kendo Dojo (NCKF, AUSKF, FIK member) and is an alumni/member of the ICU/Osawagi Kendo Dojo in Tokyo, Japan. He also serves as a board member of the Northern California Kendo Federation.

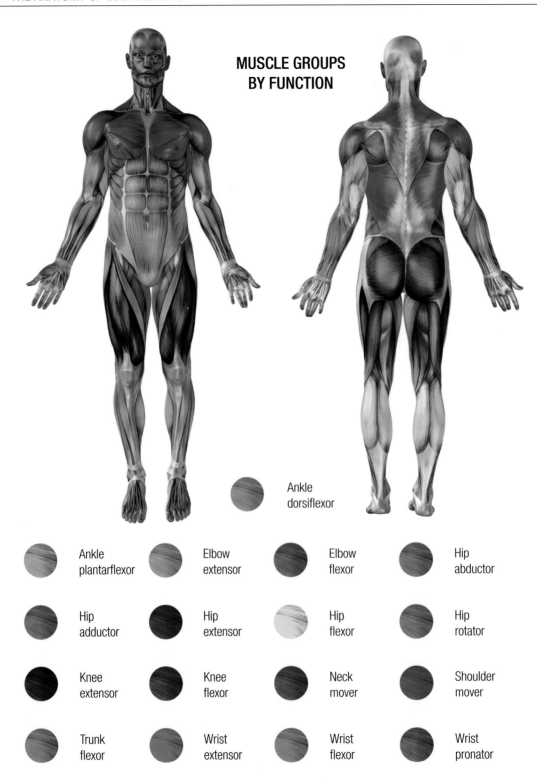

**MUSCLE GROUPS
BY FUNCTION**

Ankle
dorsiflexor

Ankle
plantarflexor

Elbow
extensor

Elbow
flexor

Hip
abductor

Hip
adductor

Hip
extensor

Hip
flexor

Hip
rotator

Knee
extensor

Knee
flexor

Neck
mover

Shoulder
mover

Trunk
flexor

Wrist
extensor

Wrist
flexor

Wrist
pronator